GMRS Radio Field Manual

The Visual Command System for FCC Licensing, Programming CTCSS/DCS Privacy Tones, Using Repeaters, and Building an Off-Grid Family Emergency Network

MorseCode Publishing

Contents

Introduction   1

The Communications Gap

    You didn't get 35 miles. You barely got to the grocery store.

    The Identity Crisis: You Are Not a "Sad Ham"

    You are a Citizen Operator.

    The Solution: The Visual Command System

    How To Use This Manual

1. The Physics of Reality   7

    Visualizing Radio Waves

       1.1 The Invisible Terrain

       1.2 The Wattage Myth

       1.3 FRS vs. GMRS: The Hybrid Confusion

       Visual Command Action Step: The "Range Expectation" Checklist

2. Licensed to Operate   16

    The 15-Minute Legal Fix

       2.1 The "No-Test" License

       2.2 Navigating the FCC Nightmare

       2.3 Your Call Sign

       Workbook Element: The "Call Sign Card"

3. Hardware Selection   25

    Buying the Right Tool

3.1 The Handheld (HT) Landscape

3.2 The Mobile Command Unit

3.3 Antenna Anatomy

Action Step: The "Gear Audit"

4. The Visual Command Programming Guide 34

Conquering CHIRP

4.1 The Driver Disaster

4.2 The "Golden Config" Spreadsheet

4.3 Creating Your Fleet

Workbook Element: The "Visual Command CHIRP Template"

Supplemental: Deep Dive on "Why the Factory Config Sucks"

Supplemental: The "Repeaterbook" Integration

Chapter Summary Checklist

5. Privacy Codes & Filters 46

The "Doorbell" Analogy

5.1 It Is NOT Encryption

5.2 CTCSS vs. DCS

5.3 Tone Squelch vs. Tone Transmit

Action Step: The "Family Tone" Setup

Appendix: The "Rosetta Stone" (Abbreviated)

6. Repeater Operations 58

Extending Your Reach

6.1 The "Split Highway" Concept

6.2 Finding Repeaters

6.3 Programming a Repeater Channel (Field Programming)

Checklist: "Repeater Pre-Check"

7. Protocol & Etiquette 68

Sounding Like an Operator

7.1 The Anti-Sad Ham Manifesto

7.2 Scripts for Anxiety-Free Comms

7.3 The NATO Phonetic Alphabet

Action Step: The "Driveway Drill"

8. The Family PACE Plan   78

SOPs for Safety

8.1 PACE Defined: The Algorithm of Survival

8.2 Kid Protocols: Managing the "Junior Operators"

8.3 The "Lost Comms" Procedure

Workbook Element: The PACE Card

9. The Mobile Build   87

Overland Install Guide

9.1 Power & Noise: The Lifeblood of Your Comms

9.2 Routing the Coax: The Arteries of the System

9.3 SWR (Standing Wave Ratio): The Engine Tune-Up

Action Step: The "Install Inspection" (The 10-Point Check)

10. Troubleshooting   98

The Visual Command Flowcharts

10.1 The "I Can't Hear You" Flowchart (Receiver Failure)

10.2 The "You Can't Hear Me" Flowchart (Transmitter Failure)

10.3 The "Radio is Hot/Smells" Flowchart (Hardware Failure)

Action Step: The "Broken Arrow" Drill

11. Advanced Operations   109

Scanning & Monitoring

11.1 Scanning Strategies: The Electronic Sentry

11.2 Dual Watch: The Two-Eared Operator

11.3 Digital/Data (The Future)

The Drill

Conclusion   119

Status - Clear

The Core Principles: A Final Review

The Challenge: The Law of Entropy

The "First Saturday" Checklist

Final Encouragement: You Are the Hub

End of Transmission

Build Your Library   126

    The Baofeng Radio Revolution

    Ham Radio Technician Class Study Guide

    Ham Radio General Class License Study Guide

Appendix: The Visual Command Toolkit   129

Quick Reference Cards

Glossary   132

Introduction

The Communications Gap

OBJECTIVE: SHIFT YOUR MINDSET from a disappointed consumer to a capable "Citizen Operator" by exposing marketing lies and rejecting the gatekeeping of hobbyists.

Key Deliverable: A rejection of the "35-Mile Lie" and adoption of the "Visual Command" mindset that values utility over theory.

You are standing in the aisle of a sporting goods store, or maybe scrolling through Amazon late at night. You're looking at a blister pack of two camo-print handheld radios. The packaging features a rugged-looking guy scaling a mountain, and in bold, jagged yellow font, it promises: **"UP TO 35-MILE RANGE."**

You buy them. You charge them. You hand one to your spouse or your kid, and you drive down the road to test this military-grade technology.

At half a mile, the signal gets fuzzy. At one mile, it sounds like bacon frying in a jet engine. At two miles - nothing. Silence. The confidence you felt in the store evaporates, replaced by the sinking realization that if this were a real emergency, you would be alone.

You didn't get 35 miles. You barely got to the grocery store.

This is the **Communications Gap**. It is the chasm between what the marketing department promised you and what physics actually allows. If you are reading this book, chances are you've felt that specific sting of disappoint-

ment. You might have tossed those radios in a junk drawer, assuming you bought "toys," or perhaps you assumed you did something wrong.

You didn't do anything wrong. You just fell for the "35-Mile Lie."

Manufacturers are technically allowed to claim that range based on a "perfect scenario" - usually involving two mountaintops with zero atmosphere, zero trees, zero buildings, and the curvature of the earth conveniently flattened out. But you don't live in a vacuum on a flat earth. You live in the suburbs, or you drive through dense pine forests, or you explore canyons.

In the real world, radio is messy. It bounces off skyscrapers, gets absorbed by wet pine needles, and dies instantly inside a metal parking garage.

This manual exists to close that gap. We are going to take the radio from a magical mystery box that disappoints you, and turn it into a reliable tool that secures your tribe.

The Identity Crisis: You Are Not a "Sad Ham"

Before we fix your radio, we need to fix your mindset.

If you have ever tried to research radio online, you likely stumbled into a forum or a Facebook group populated by what we affectionately call "Sad Hams."

The Sad Ham is a gatekeeper. He has been into Amateur Radio (Ham) since 1978. He speaks entirely in acronyms. He is obsessed with electrical theory, soldering irons, and obscure regulations. If you ask a simple question like, *"Which radio is best for my Jeep?"* the Sad Ham will lecture you. He will tell you that you're too stupid to operate a radio, that your equipment is "Chinese junk," and that you need to study for a difficult technical exam to earn the privilege of pushing a button.

The Sad Ham treats communication as a **hobby**. He wants to build the radio. He wants to talk to a stranger in Slovenia about the weather using a wire strung up in his backyard.

That is not you.

You are a Citizen Operator.

For him, the radio is the destination. For you, the radio is just the vehicle. It is a tool that supports your actual hobbies and responsibilities.

- You are the **Tactical Dad** ensuring your family can coordinate during a power outage or natural disaster.

- You are the **Overland Adventurer** who needs to tell the truck behind you that there is a washout around the blind curve.

- You are the **Community Protector** organizing a neighborhood watch.

You don't care about the physics of wave propagation in the ionosphere. You care about: *If I press this button, will my wife hear me?*

You want utility, not theory. You want results, not a science project. This book is written specifically for the Citizen Operator. We are explicitly rejecting the gatekeeping culture of the radio world. We don't care if your radio cost $30 or $300; we care if you know how to use it when the cell towers go down.

The Solution: The Visual Command System

Why is radio so confusing? Because it is invisible.

You can't see radio waves. You can't see why a signal hits your friend standing on a hill but misses your friend standing in the valley. When things are invisible, they feel like magic. And when magic stops working, we feel helpless.

To become a competent operator, we have to make the invisible **visible**.

This book utilizes the **Visual Command System**. We are going to bypass the complex electrical engineering formulas (which you will never use in the field) and replace them with physical, visual analogies that your brain can instantly grasp.

- We won't talk about "electromotive force"; we will talk about **Water Pressure**.

- We won't talk about "RF propagation shadowing"; we will talk about **Flashlight Beams**.

- We won't talk about "squelch hysteresis"; we will talk about a **Gatekeeper**.

If you can understand how a garden hose nozzle focuses water, you can understand how to double your range with a better antenna. If you can understand how a flashlight casts a shadow behind a wall, you can understand why your radio doesn't work in the city.

This system relies on "If-This-Then-That" logic. We are stripping away the fluff to give you a decision tree.

- *Static on the line?* Do this.

- *Can't hear your buddy?* Move here.

- *Radio beeping loudly?* Turn this off.

By the end of this manual, you won't just be a person holding a plastic box. You will be a Commander of your immediate environment. You will look at the terrain - the hills, the buildings, the trees - and you will visually "see" where your signal is going. You will operate with confidence because you understand the mechanical reality of what is happening in the air around you.

How To Use This Manual

This is not a textbook. Do not put this on your bookshelf between your college history books and that novel you haven't read yet.

This is a **Field Manual**.

It is designed to live in your glovebox, your center console, or the top pouch of your Go-Bag. It is designed to get dog-eared, stained with coffee, and marked up with a sharpie. The chapters are short for a reason. The diagrams are bold for a reason. When you are stuck on a trail in the rain and your comms go down, you don't have time to read a dissertation. You need an answer.

The Go-Bag Size

You will notice we have kept the formatting tight. We have stripped out the history of the FCC and the biography of Marconi. We focus entirely on GMRS (General Mobile Radio Service) because it is the "Goldilocks" of radio for the regular person - more powerful than the toy blister-pack radios (FRS), but requiring no tests or technical exams, unlike Ham radio.

The Icons

To help you scan this book quickly when you are in the field, we use a system of visual icons:

THE SAFETY ALERT When you see this icon, pay attention. This indicates a warning. Usually, this means "Don't do this or you will fry your radio" or "Don't do this or you will interfere with emergency services." These are the guardrails. Stay between them, and you can't mess up.

FIELD TACTICS This icon marks a force multiplier. These are the tips and tricks that instantly boost your range or clarity. These are the hacks that separate the rookies from the pros. When you see this, you're about to learn a shortcut that actually works.

MENTAL MODEL This icon introduces a visualization tool. This is where we turn a complex invisible concept into a simple picture. Read these sections carefully; they are the foundation of the Visual Command System.

You have already taken the first step. You realized that a cell phone is a fragile leash. You realized that in a real emergency, or even just deep in the woods, the grid does not care about you.

You bought the radio. Now, let's learn how to actually use it. Turn the page, and let's get to work.

Jared (WSLC230)

Jared@morsecodepublishing.com

Basic Structure of a GMRS Radio

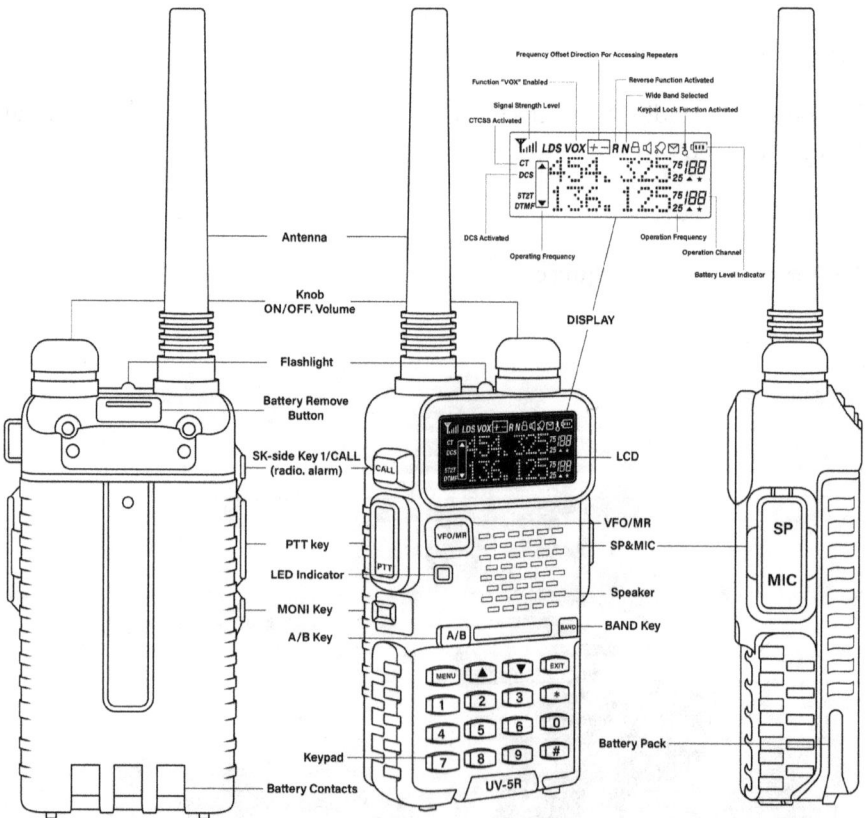

Frequency Offset Direction For Accessing Repeaters

Function "VOX" Enabled

Signal Strength Level

Reverse Function Activated

Wide Band Selected

Keypad Lock Function Activated

CTCSS Activated

DCS Activated

Operating Frequency

Operation Frequency

Operation Channel

Battery Level Indicator

Antenna

Knob
ON/OFF. Volume

DISPLAY

Flashlight

Battery Remove
Button

SK-side Key 1/CALL
(radio, alarm)

LCD

VFO/MR

PTT key

SP&MIC

LED Indicator

Speaker

MONI Key

A/B Key

BAND Key

Battery Pack

Keypad

Battery Contacts

UV-5R

SP

MIC

Chapter 1

The Physics of Reality

Visualizing Radio Waves

OBJECTIVE: REPLACE INVISIBLE MAGIC with physical intuition regarding how terrain, obstacles, and antenna height dictate signal success.

Key Deliverable: The "Range Expectation Checklist"—a tool to instantly predict signal failure based on your environment (Urban Canyon vs. Open Desert).

Most people try to solve their communication problems with their credit card.

They buy a more expensive radio. They buy a "high-gain" antenna that looks like a tactical whip. They buy a 50-watt mobile unit to replace their 5-watt handheld, assuming that ten times the power equals ten times the range.

Then they go out into the field, press the button, and fail.

They fail because they are fighting physics, and physics always wins. You cannot buy your way out of the laws of nature. However, you can *outsmart* them.

Welcome to the foundation of the **Visual Command System**. In this chapter, we are going to stop thinking about radio waves as invisible, magical forces. We are going to visualize them as physical objects. Once you can "see" the radio waves leaving your antenna, you will intuitively know where to stand, where to drive, and when to give up on a bad position.

We are not going to do math. We are going to do geometry.

1.1 The Invisible Terrain

To understand GMRS (General Mobile Radio Service), you have to understand where it lives on the spectrum. GMRS operates in the UHF (Ultra High Frequency) band, specifically around 462 MHz and 467 MHz.

Unless you are an electrical engineer, those numbers mean nothing to you. So let's replace them with a mental model.

Mental Model: The Flashlight vs. The Bouncing Ball

Imagine you are holding two different tools in a dark room full of furniture.

1. The CB Radio (Low Frequency): Think of a CB radio (27 MHz) like a **super-bouncy rubber ball**. You throw it (transmit). It hits a wall, it bounces. It hits a ceiling, it bounces. It hits the ground, it bounces. In the right conditions (atmospheric skip), you can throw that ball, and it might bounce off the sky (ionosphere) and land in another state. It's chaotic, it's long-range, but it's unreliable. It goes around corners easily, but it has no precision.

2. The GMRS Radio (UHF Frequency): Think of your GMRS radio like a **tactical flashlight**. When you press the PTT (Push-to-Talk) button, you are turning on a beam of light.

- **Does light go through a mountain?** No. It hits the rock and stops.

- **Does light go around the curve of the earth?** No. It travels in a straight line until it hits the horizon.

- **Does light go through a thick forest?** Sort of. Some beams get through the gaps in the leaves, but a lot of it gets absorbed or scattered.

This is the most critical rule of GMRS: **Line of Sight.**

If your antenna cannot "see" the other antenna (optically speaking), you probably won't talk to them. The radio waves travel in straight lines, just like light. If there is a massive granite ridge between you and your convoy, you are in the dark. No amount of wattage will blast through that rock.

The "Urban Canyon": Why GMRS Beats CB

If GMRS is just a "flashlight" that gets blocked by obstacles, why do we use it? Why not use the "bouncy ball" CB radio?

Because we don't live on a flat desert. We live in the **Urban Canyon**.

The Urban Canyon is any environment with man-made structures - cities, suburbs, or even dense industrial parks. In this environment, the "bouncy ball" of the CB radio is too big. The wavelength is 11 meters long (about 36 feet). Trying to get a 36-foot wave into a parking garage or through a window is like trying to drive a semi-truck through a doggy door. It doesn't fit.

GMRS waves (UHF) are short - about 70 centimeters (2 feet).

This is small enough to fit through windows. It's small enough to bounce down a hallway. It's small enough to punch through drywall and standard suburban housing materials (with one major exception: **Stucco**, which often contains a metal wire mesh that kills signals).

 Field Tactic: The Window Breach If you are inside a building and can't reach your team, move to a window. Glass is transparent to UHF radio waves, just like it is to light. But **watch out for insect screens**. If your window has a metal screen, it acts as a Faraday cage and blocks the signal. You must open the screen to let the 'beam' escape. If you are in a basement? You are dead. The earth eats radio waves. Get high, get to a window.

Diagramming Line of Sight (LOS)

Picture two people standing on a flat prairie.

- **Person A** is 6 feet tall.

- **Person B** is 6 feet tall.

Due to the curvature of the earth, the horizon is only about 3 miles away. If Person A and Person B stand 7 miles apart on perfectly flat ground, they cannot see each other. The earth creates a hump between them.

Since GMRS is a "flashlight," the beam from Person A's radio hits the dirt hump of the earth before it reaches Person B.

This is why the "35-Mile Range" on the box is a lie. That 35-mile claim assumes Person A is on top of Mount Everest and Person B is in a valley, with absolutely nothing in between. In the real world, the earth curves.

The Solution? You have to elevate the flashlight. If Person A climbs a water tower, he can now "see" over the curvature of the earth. His radio beam clears the obstacles. This leads us directly to the single most misunderstood concept in radio: Power vs. Height.

1.2 The Wattage Myth

If you hang out in truck stops or 4x4 forums, you will hear guys bragging about their wattage. *"I've got a 50-watt mobile unit in the dash." "I hacked my Baofeng to push 8 watts."*

They treat wattage like horsepower in a truck. They think 50 watts is "stronger" and will therefore push the signal further, like a battering ram smashing through obstacles.

This is the Wattage Myth.

Mental Model: The Water Hose

Imagine your radio is a water hose.

- **Wattage** is the **Water Pressure** (PSI).

- **The Antenna** is the **Nozzle**.

If you have a hose pointing straight at a brick wall (a mountain), you can turn the pressure up to 1,000 PSI. What happens? You just splash more water back on yourself. You don't punch through the wall.

Adding wattage does not help you penetrate obstacles. It only helps you push the signal further *if the path is already clear*.

The Inverse Square Law (The 'Napkin Math' Version)

Here is the hard truth that makes grown men cry: **To double your distance, you have to quadruple your power.**

Actually, it's often worse than that. In real-world conditions, the returns diminish incredibly fast.

Let's look at the difference between a standard handheld and a high-power mobile unit:

- **5 Watts (Handheld):** This is your baseline. Let's say this gets you 2 miles in the suburbs.

- **50 Watts (Mobile Unit):** This is 10 times the power. Do you get 20 miles? **No.**

You might get 3 miles. Maybe 4.

You have burned 10x the battery power for a measly 20% gain in distance.

Why? Because you are still fighting the curvature of the earth and the trees. You are just "yelling louder" at the trees. The trees don't care.

When DOES wattage matter? Wattage matters for **clarity**, not necessarily distance. Think of a crowded room (static/interference).

- A 5-watt radio is speaking in a normal voice.

- A 50-watt radio is shouting through a megaphone.

If you are within range, the 50-watt radio will sound crisp, loud, and authoritative. It will burn through the background static ("noise floor"). But it will not magically bend around a mountain.

The "Height is Might" Principle

If you have $200 to spend on improving your comms, and you have to choose between a more powerful radio or a better antenna location, choose the antenna every single time.

Height is the only thing that defeats geometry.

Remember the flashlight.

- **Scenario A:** You stand on the ground with a 10-million-candlepower searchlight (High Wattage). You shine it forward. It hits the trees 100 yards away. Range: 100 yards.

- **Scenario B:** You climb a 50-foot tower with a cheap disposable penlight (Low Wattage). You shine it forward. You can see for miles.

Field Tactic: The "Roof Rack" Rule Never use a handheld radio inside a vehicle if you can avoid it. The metal body of your car is a "Faraday Cage" - it blocks radio waves. It keeps the signal *in*.

- **Bad:** Holding a 5-watt radio in the driver's seat. (Effective power escaping the car: maybe 1 watt).

- **Good:** Connecting that handheld to a magnetic mount antenna on the roof.

- **Best:** A dedicated mobile unit with a hard-mounted antenna on the roof rack.

By moving the antenna from your lap to the roof, you gain about 4 to 6 feet of elevation. In the radio world, that is huge. You have cleared the engine block, the dashboard, and the heads of your passengers. You have given your "flashlight" a clear view of the horizon.

The Golden Rule: A 5-watt radio on a mountaintop beats a 50-watt radio in a valley. Every. Single. Time.

1.3 FRS vs. GMRS: The Hybrid Confusion

Now that we understand the physics, we have to clear up the legal and technical mess of the channels.

You will often hear terms like FRS (Family Radio Service) and GMRS (General Mobile Radio Service) used interchangeably. This is because they share the same "driveway."

Years ago, the FCC made a decision that confused everyone. They allowed FRS (the cheap, blister-pack radios you buy at Walmart) to operate on the exact same frequencies as GMRS (the prosumer, licensed gear).

This means your high-speed, 50-watt, waterproof, overland GMRS mobile unit can talk to your nephew's $15 Spider-Man walkie-talkie.

But there are rules.

The "Bubble Pack" Chart

Think of the radio spectrum like a highway with 22 lanes (Channels 1 through 22).

Channels 1–7 (The "Low" Shared Lanes)

- **Who is here?** Everyone. FRS and GMRS.

- **Power Limit:** limited to 5 watts for GMRS (formerly less, but rules modernized).

- **Vibe:** Crowded. This is where you hear kids playing tag and people parking RVs.

Channels 8–14 (The "Weak" FRS-Only Lanes)

- **Who is here?** technically FRS only, with very strict power limits (0.5 watts).

- **GMRS Restriction:** Most proper GMRS radios are hard-locked to transmit at very low power here, or they skip these channels entirely.

- **Vibe:** Silence or static. Because the power is so low (0.5 watts is like a whisper), these channels are useless for anything beyond "across the campsite" comms. **Avoid these for tactical use.**

Channels 15–22 (The "High" GMRS Power Lanes)

- **Who is here?** The Big Dogs.

- **Power Limit:** 50 Watts.

- **Vibe:** This is where we live. This is where you get the range. If you are running a convoy, a neighborhood watch, or a farm operation, you park your comms here.

SAFETY ALERT: The Wide vs. Narrow Band Trap High-quality GMRS radios use "Wide Band" FM audio. Cheap FRS radios use "Narrow Band." If you talk from a GMRS radio (Wide) to an FRS radio (Narrow), you will sound incredibly loud and distorted to them - like you are swallowing the microphone. If they talk to you, they will sound quiet and thin.

- **Action:** If you know your group has a mix of cheap and expensive radios, go into your GMRS radio settings and switch your bandwidth to **NARROW (NFM)**. This levels the playing field. Remember: The cheap FRS radio is **locked** to Narrow band. The user cannot change it. You are the 'Command' station; it is your job to adapt your equipment to hear them, not the other way around.

Why "Repeater Channels" Are Different

You might see channels 15-22 listed, and then another set of channels called "RPTR 15" or "Rep 15."

These are the **Repeater Inputs**. We will cover repeaters in depth in Chapter 5, but for now, visualize a repeater as a "Range Extender Tower."

- **Channels 1-22:** You talk directly to another radio (Simplex). Like tossing a ball to a friend.

- **Repeater Channels:** You talk to a Machine on a mountain, and the Machine screams your message down to the valley (Duplex). Like throwing a ball at a backboard.

Only GMRS radios can access repeaters. FRS radios cannot. This is the main reason you paid for the GMRS license. It unlocks the "Cheat Mode" of using someone else's tower to get 50 miles of range.

Visual Command Action Step: The "Range Expectation" Checklist

Stop guessing. Stop hoping. Use this checklist to set realistic expectations for your team before you step out the door.

This table assumes you are using a standard **5-Watt Handheld Radio** (the most common scenario).

THE TERRAIN	VISUAL ANALOGY	REALISTIC RANGE	WHY?
Inside a Vehicle	The Faraday Cage	0.5 – 1 Mile	The metal roof blocks the signal "flashlight" from escaping.
Suburban Neighborhood	The Urban Canyon	1 – 2 Miles	Houses, fences, and trees scatter the beam.
Dense Forest (Flat)	The Sponge	0.5 – 1.5 Miles	Pine needles and water-filled trunks absorb UHF energy.
Open Water / Desert	The Billiard Table	4 – 6 Miles	No obstacles to block the beam. (Earth curvature eventually cuts it).
Hilltop to Valley	The Sniper's Nest	5 – 25+ Miles	You have defeated the earth's curvature. Pure Line of Sight.

Your Homework: Go outside. Look at your environment. Don't look at it like a landscape; look at it like a geometry problem. Where are the lines of sight? Where are the shadows?

If you are in a valley, **move**. If you are behind a building, **move**. If you are frustrated, **get your antenna higher**.

That is the physics of reality. Now that you understand the playing field, let's look at the equipment you need to dominate it.

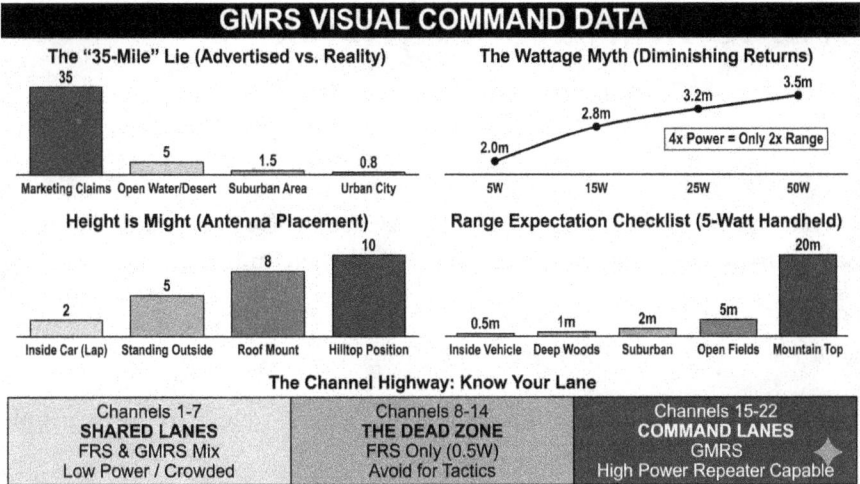

GMRS VISUAL COMMAND DATA

The "35-Mile" Lie (Advertised vs. Reality)

Marketing Claims	Open Water/Desert	Suburban Area	Urban City
35	5	1.5	0.8

The Wattage Myth (Diminishing Returns)

5W	15W	25W	50W
2.0m	2.8m	3.2m	3.5m

4x Power = Only 2x Range

Height is Might (Antenna Placement)

Inside Car (Lap)	Standing Outside	Roof Mount	Hilltop Position
2	5	8	10

Range Expectation Checklist (5-Watt Handheld)

Inside Vehicle	Deep Woods	Suburban	Open Fields	Mountain Top
0.5m	1m	2m	5m	20m

The Channel Highway: Know Your Lane

Channels 1-7 **SHARED LANES** FRS & GMRS Mix Low Power / Crowded	Channels 8-14 **THE DEAD ZONE** FRS Only (0.5W) Avoid for Tactics	Channels 15-22 **COMMAND LANES** GMRS High Power Repeater Capable

Chapter 2

Licensed to Operate

The 15-Minute Legal Fix

OBJECTIVE: DEMYSTIFY THE LEGAL bureaucracy to legitimize the entire family unit without fear of exams, Morse code, or "Sad Ham" judgment.

Key Deliverable: The "Family Umbrella" strategy that covers your whole tribe with a single $35 fee, plus a printable "Call Sign Card" for the glovebox.

There is a dirty secret in the off-road and prepping communities: **Most people are operating illegally.**

You see it all the time. A guy buys a $30 radio on Amazon, unboxes it, and starts blasting out transmissions on GMRS frequencies without a license. He assumes that because he bought the radio at a store, he is allowed to use it. Or, he assumes that getting a license involves sitting in a high school cafeteria on a Saturday morning, taking a calculus exam proctored by three guys in ham radio vests who are judging his lack of knowledge about Ohm's Law.

This fear - the fear of the "Test" - is the single biggest barrier to entry for the Citizen Operator.

Let's kill that fear right now.

There is no test. There is no exam. There is no Morse code. There is no interview.

Getting a GMRS license is exactly like getting a fishing license. You go to a government website, you fill out a form that says you aren't a foreign spy, you pay a fee (currently $35), and they email you a PDF.

That's it.

The entire process takes about 15 minutes. It covers your entire family for 10 years. It gives you legal access to powerful repeaters that can triple your range.

If you can order a pizza online, you can get a GMRS license. This chapter is your walkthrough. We are going to navigate the government bureaucracy together, dodge the confusing acronyms, and get you legitimate.

2.1 The "No-Test" License

Why does the "Sad Ham" (Amateur Radio Operator) have to take a test, but you don't?

It comes down to **responsibility**.

- **Ham Radio:** The operator is allowed to build their own radios, modify electronics, and use massive power levels on frequencies that can bounce around the world. Because they can build dangerous things, they need to prove they understand the physics.

- **GMRS:** The radios are certified appliances. You cannot build a GMRS radio; you can only buy one that has been FCC-approved. Since the equipment is "idiot-proofed" by the manufacturer, you don't need to know the physics. You just need to know the rules.

Therefore, the GMRS license is not a "Certificate of Competence." It is simply a tax. It is a "Pay-to-Play" fee that registers you in the database.

The "Family Umbrella" Rule

This is the killer feature of GMRS.

In Ham radio, the license belongs to the *individual*. If you have a license, your wife cannot talk on the radio unless she gets her own license (or you stand right next to her as a "control operator").

In GMRS, the license belongs to the **System**.

When you pay your $35, you receive a Call Sign (e.g., WRXP-555). The FCC regulations (47 CFR § 95.1705) state that this single license covers you **and your immediate family members.**

This means you do not need to buy a $35 license for your wife. You do not need one for your son. You do not need one for your daughter. One license covers the tribe.

Mental Model: The Family Flowchart

Who exactly is "family"? The FCC definition is surprisingly generous.

Picture a family tree. **YOU (The License Holder)** are the trunk.

Who is covered?

- **Spouse:** Yes.

- **Children / Step-children:** Yes.

- **Grandchildren:** Yes.

- **Parents / Step-parents:** Yes.

- **Grandparents:** Yes.

- **Siblings:** Yes.

- **Aunts/Uncles/Nieces/Nephews:** Yes.

- **In-Laws:** Yes (including Father/Mother-in-law, Brother/Sister-in-law, and Son/Daughter-in-law).

Who is NOT covered?

- **Cousins:** No. (This is the most common mistake. Cousin Eddie needs his own license).

- **Employees:** No. If you run a landscaping business, you cannot use a personal GMRS license for your crew. That requires a Commercial Business License (Part 90), which is a totally different beast.

- **Friends/Neighbors:** No. Your buddy in the truck behind you needs his own license.

Field Tactic: The "Guest Operator" Loophole "But wait," you ask. "I'm taking a friend camping. I hand him a spare radio. Is he operating illegally?"

Technically? Yes. However, the rules allow for emergency messages. If safety is at risk, anyone can use any radio. But for general chit-chat ("Hey, grab me a beer"), your friend is technically in violation.

The Solution: Tell your friend to spend the $35. If your group has 3 heads of households (you, your brother, your best friend), you need 3 licenses. That covers all the wives, kids, and grandparents in the convoy. It is a small price to pay for a 10-year permit.

2.2 Navigating the FCC Nightmare

The hardest part of GMRS isn't the radio; it's the website.

The Federal Communications Commission (FCC) uses a system called **CORES** (Commission Registration System) and **ULS** (Universal Licensing System). These websites appear to have been built in 1998 and never updated. They are clunky, they time out, and they use confusing terminology.

Do not let the bad UI (User Interface) defeat you. Follow this "If-This-Then-That" walkthrough.

Phase 1: Get Your FRN (FCC Registration Number)

Before you can apply for a license, you need a customer ID number. This is your FRN.

1. Go to the **FCC CORES** website.

2. Click **"Register"** to create a username (use your email).

3. Once logged in, select **"Register New FRN"**.

4. **Select "An Individual"** (unless you are a corporation, which you aren't).

5. **Contact Address:** Yes, you must use your real address. Yes, this becomes public record.

   ○ *Privacy Note:* If you are concerned about OPSEC (Operational Security) and don't want your home address searchable in the FCC database, use a P.O. Box. The FCC does not care if you live there, but they must be able to mail you enforcement letters.

6. **Social Security Number:** You must provide it. This is to ensure you aren't a felon avoiding federal fees.

7. **Submit.** You will instantly get a 10-digit number. **Write this down.** This is your FRN.

Phase 2: The Application (The "ZA" Code)

Now that you have an FRN, you can buy the license.

1. Log in to the **FCC ULS License Manager**.

2. On the left menu, click **"Apply for a New License"**.

3. **The Dropdown Menu Trap:** You will see a list of radio services with confusing acronyms. You are looking for **"ZA - General Mobile Radio (GMRS)"**. It is usually at the very bottom of the list.

   ○ *Note:* Do not click "Amateur" (Ham). Do not click "Land Mobile". Click **ZA**.

4. **Applicant Questions:**

   ○ *Has the applicant been convicted of a felony?* Answer honestly. (A felony does not automatically disqualify you, but it triggers a review).

   ○ *Do you agree to the rules?* Yes.

5. **Summary:** Verify your address and name.

6. **"Certify"**: This is your digital signature. Type your name.

Phase 3: The Payment Loop (The Final Boss)

This is where 50% of people fail. After you click "Submit Application," the ULS system will say "Application Submitted." **You are not done.**

You must now switch *back* to the **CORES** system to pay. The two systems do not talk to each other well.

1. Look for a button that says **"Log in to CORES to Pay Fees"**.

2. If you miss it, just log back into CORES separately.

3. You should see a "Red Light Status" or "Pending Payment" notification.

4. Select the pending application.

5. Pay the **$35**. (Credit card is best).

Still having issues finding it? Follow this path on the website: FCC website > FCC Registration page > Manage Existing FRNs > FRN Financial page

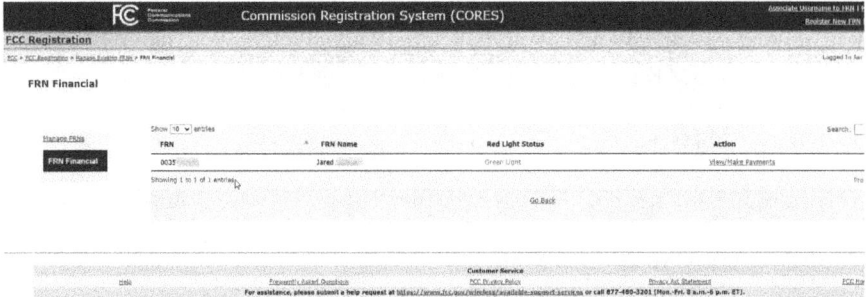

Phase 4: The Waiting Game

You will not get your license instantly. It usually takes **24 to 48 business hours**. Check your email (including spam) for a message from the FCC with a link to "Download Authorization."

 SAFETY ALERT: Do Not Transmit Yet Do not pick up that radio and start talking while your application is 'Pending.' You are not legal until the FCC grants the license and issues the Call Sign. Wait for the email.

 Field Tactic: The Reference Copy Trap When you finally log in to download your license, you will see two options:

1. "Reference Copy" (Watermarked "Reference Only")

2. "Official Authorization"

Download the **Official Authorization**. Print it out. Put one copy in your glovebox. Put one copy in your Go-Bag. Save a screenshot of it on your phone in a "Important Docs" album. You will likely never be asked for it by law enforcement, but repeater owners (the guys who own the big towers) will often ask for your Call Sign and proof of license before letting you use their system.

2.3 Your Call Sign

Congratulations. You are now **WRXP-555** (or whatever jumble of letters and numbers they assigned you).

This is your **Call Sign**. It is your legal identity on the airwaves.

In the movies, people use "Handles" like *Rubber Duck* or *Eagle Eye*. In the real world of GMRS, we use Call Signs.

When to Identify

The rules (47 CFR § 95.1751) are simple, but people overcomplicate them.

You must state your Call Sign:

1. **Every 15 minutes** during a conversation.

2. **At the end** of a conversation.

You DO NOT need to state it:

- At the very beginning (though it is polite).

- In every single sentence.

Example of a Proper Exchange:

- **Start:** *"Unit 1 to Unit 2, radio check."*

- **Middle:** (You talk about the trail, the weather, the lunch stop for 10 minutes).

- **End:** *"Unit 1 clear. WRXP-555."*

That's it. You just have to tag the end of the chain. If you are talking for an hour, just throw your call sign out whenever there is a lull. *"Hey, just ID-ing, this is WRXP-555."*

The "Unit Number" System (Tactical Clarity)

Your Call Sign is a mouthful. *"Whiskey Romeo X-ray Papa Five Five Five."* Trying to say that every time you want to tell your wife to turn left is annoying.

The Solution: Use **Unit Numbers** for tactical comms, and **Call Signs** for legal ID.

Assign everyone in your family/group a tactical name:

- **Dad:** "Unit 1" (or "Lead")

- **Mom:** "Unit 2"

- **Kids:** "Base" (if at camp) or "Unit 3".

The Conversation: *"Unit 1 to Unit 2, slow down, big rock on the left." "Unit 2, copy."* (Conversation continues...) *"Unit 1, we are stopping for gas." "Unit 2, copy. WRYZ-123 clear."* (Mom drops the legal ID at the end to cover the whole family).

Since the license covers the whole family, **only one person needs to ID** to make the whole transmission legal. If Dad says the Call Sign, Mom is covered.

The Myth of "The FCC Van"

Let's address the paranoia. If you forget to ID? If you say your Call Sign wrong? Will a black van with antennas skid into your driveway and arrest you?

No.

The FCC Enforcement Bureau is underfunded and overworked. They are hunting two things:

1. **Commercial Interference:** A construction company using police frequencies and blocking 911 calls.

2. **Malicious Interference:** Someone setting up a high-power transmitter specifically to jam a repeater or scream profanity for days on end.

They are not hunting "Tactical Dad" who forgot to ID his call sign while hiking.

However... While the *government* might not be listening, the **Sad Hams** are. The GMRS community is self-policing. If you hop on a public repeater and start acting like an idiot - playing music, not using a call sign, burping into the mic - the owner of that repeater can and will block you. They can remotely turn off the repeater or ban your specific radio ID if you are using digital modes.

Respect the Commons. We share these airwaves. Using your Call Sign isn't just a legal requirement; it's a signal to everyone else that you are a "Citizen Operator" who knows the rules, not a "Blister Pack Rookie" who is just making noise. It buys you respect.

Workbook Element: The "Call Sign Card"

Do not rely on your memory. In an emergency, your adrenaline will be high, and you will forget whether you are "WRXP" or "WXRP".

Print this template, fill it out with a Sharpie, and tape it to the back of every radio in your fleet. *Scan to Download Printable Copies.*

Chapter 3

Hardware Selection

Buying the Right Tool

OBJECTIVE: CUT THROUGH "GEAR Paralysis" to select hardware based on Mission Profile (Burner vs. Operator) rather than confusing marketing specs.

Key Deliverable: A "Mission-Ready" hardware loadout that eliminates weak points like bad antennas, heat traps, and incompatible connectors.

If you ask a forum full of radio enthusiasts what radio you should buy, you will get 50 different answers, and 49 of them will be wrong for you.

One guy will tell you to buy a $600 commercial Motorola that requires proprietary software from 1998 to program. Another will tell you to buy a $20 Baofeng because "it's just as good" (it isn't). A third will tell you to build your own radio out of a tuna can and copper wire.

This is **Gear Paralysis**.

In the Tactical Dad and Overlanding world, we love gear. We love researching specs. But with radio, the specs on the box rarely tell the full story. A radio that promises "50 Watts" might have a receiver so sensitive that it goes deaf the moment you drive near a Taco Bell drive-thru.

In this chapter, we are going to cut through the marketing noise. We are going to categorize hardware not by price, but by **Mission Profile**.

Are you equipping a neighborhood watch? Are you outfitting a Jeep for a cross-country trip? Are you handing a radio to a toddler?

We need the right tool for the job.

3.1 The Handheld (HT) Landscape

The "HT" (Handie-Talkie or Handheld Transceiver) is the gateway drug of GMRS. It is portable, battery-powered, and usually affordable.

However, the market is flooded with junk. To navigate this, we divide handhelds into two distinct tiers: The **"Burner" Tier** and the **"Operator" Tier**.

Tier 1: The "Burner" (Baofeng & Clones)

- **Price:** $25 - $40

- **Typical Models:** Baofeng UV-5G, UV-9G, Retevis RA85.

- **The Vibe:** Disposable.

Everyone owns a Baofeng. They are the Harbor Freight tools of radios - cheap, they get the job done for a while, and you won't cry if you drop one in a mud puddle.

The "Glass Jaw" Problem (Front-End Overload) The biggest weakness of cheap radios is their receiver chip. They use a "Direct Conversion" receiver, which is like having ears that cannot filter sound.

- **Scenario:** You are deep in the woods. The radio works great.

- **Scenario:** You drive up to a hilltop near a city to get a better signal. Suddenly, your radio goes silent. Or it starts screeching static. You can't hear your buddy 100 yards away.

Why? Because there is a cell phone tower or a paging transmitter on that same hill. The cheap radio chip gets overwhelmed by *all* the RF energy nearby, not just the GMRS signal. It gets "blinded" by the noise. It's like trying to listen to a whisper while standing next to a jet engine.

The Verdict: Buy these as backups. Buy them to hand out to friends who didn't bring a radio. Buy them for your kids. But do not trust your life to them.

Tier 2: The "Operator" (Wouxun & High-End)

- **Price:** $100 - $250

- **Typical Models:** Wouxun KG-935G, KG-S88G, Rocky Talkie 5 Watt.

- **The Vibe:** Reliable Tool.

These radios often use **Superheterodyne** receivers (or very high-quality filtering chips). This is the "Bouncer" at the door. The radio ignores the cell towers and paging systems and only lets the specific GMRS frequency through. You can stand right next to a radio tower and still hear a weak signal from your convoy.

The Verdict: This is your primary carry. It survives drops, it filters noise, and it works when things get loud.

The Essential Features Checklist

Do not buy a handheld radio in 2025 unless it passes this 3-point inspection.

1. USB-C Charging (Non-Negotiable) Old-school radios require a "cradle" charger that plugs into a wall. This is a logistical nightmare in the field.

- *The Requirement:* You must be able to charge the battery directly via USB-C.

- *Why:* Because you already have a USB-C cable in your truck, in your solar bank, and in your laptop bag. If you lose the charging cradle for an old radio, that radio is a brick. If you have USB-C, power is everywhere.

2. Removable Antenna (SMA Connector) Some "blister pack" radios have fixed antennas glued to the body. Avoid these.

- *The Requirement:* You must be able to unscrew the antenna.

- *Why:* The stock antenna is trash. You need the ability to swap it for a high-gain whip or connect the radio to a roof-mount antenna cable.

3. IP Rating (The "Dunk" Test) Check the "IP" (Ingress Protection) code.

- **IP54:** Splash proof. Okay for light rain. Dust protected. (Minimum Standard).

- **IP67:** Submersible. You can drop it in a creek for 30 minutes.

- *Reality Check:* If you are an overlander, IP54 is fine. If you are a kayaker or a duck hunter, get IP67.

3.2 The Mobile Command Unit

Handhelds are great for spotters and hiking. But if you are in a vehicle, a handheld is the wrong tool.

We discussed the **Faraday Cage** effect in Chapter 1. The metal body of your vehicle blocks radio waves. To get a signal out, you need an external antenna. You *could* connect your handheld to an external antenna, but that leaves you with a cable dangling across your steering wheel and a battery that dies in 8 hours.

Enter the **Mobile Unit**.

These are 20-watt or 50-watt boxes hard-wired to your car battery. They never run out of power. They push 10x the wattage of a handheld.

The Problem: "Where do I mount this brick?"

In a 1995 Ford Bronco, you had plenty of dashboard space to screw in a CB radio. In a modern Toyota Tundra or Jeep Wrangler, every square inch of the dash is covered in touchscreens, airbags, and leather. There is nowhere to mount a radio box.

The Solution: "All-in-the-Handset" Controls

This form factor has taken over the market (e.g., Midland MXT275, MXT575, Wouxun KG-1000G Plus).

How it works:

1. **The Black Box:** The actual radio body is a blind black box. You hide this under the driver's seat, behind the glovebox, or in the center console.

2. **The Mic:** The microphone contains the screen, the volume knob, the channel selector, and the speaker.

3. **The Connection:** You run one ethernet-style cable from the hidden box to the dashboard.

Why this rules the world:

- **Stealth:** When you park at the trailhead, you unplug the mic and lock it in the glovebox. To a thief looking through the window, your truck looks stock. There is no expensive electronics box visible.

- **Ergonomics:** You don't have to reach across the dash to change channels. The controls are in your hand.

 SAFETY ALERT: The 50-Watt Heat Problem If you buy a high-power (50W) unit, remember that efficiency is never 100%. A lot of that power turns into **Heat**.

- *Do not* mount a 50W radio body inside a sealed center console wrapped in foam. It will overheat and shut down.

- *Do not* zip-tie it directly to a heater vent.

- Give the heatsink fins room to breathe. Under a seat is fine for daily drivers, but if you ford rivers in a Jeep, mount it high (behind the glovebox) to avoid drowning the radio in deep water crossings.

3.3 Antenna Anatomy

The antenna is 80% of your performance. The radio is just the engine; the antenna is the tires. You can have a Ferrari engine (50W radio), but if you have bald tires (bad antenna), you aren't going anywhere.

Visual Guide: The Three Shapes

1. The "Rubber Duck" (Stock Handheld)

- *Shape:* Short, stubby, flexible.

- *Physics:* It is a coiled spring inside rubber. It is physically shorter than the radio wave requires, so it uses a coil to "trick" the radio.

- *Performance:* **Poor.** It acts like a resistor. It absorbs signal. It is designed for convenience, not range.

- *Use Case:* Clipping the radio to your belt while walking around camp.

2. The "Ghost" (NMO Low Profile)

- *Shape:* A cylinder, about 3 inches tall. Looks like a soup can or a salt shaker.

- *Physics:* Unity Gain (no magnification). It radiates the signal in a perfect sphere (like a lantern).

- *Performance:* **Reliable.** It doesn't snag on tree branches. It fits in parking garages. It handles high power.

- *Use Case:* The daily driver. Because it is only 3 inches tall, you never have to take it off to enter your home garage or a parking structure. It is the 'Set it and Forget it' option.

3. The "Whip" (High Gain)

- *Shape:* A thin steel rod, usually 32 inches or longer. Sometimes has a "coil" in the middle.

- *Physics:* High Gain (3dB to 6dB). It squashes the signal sphere into a flat pancake. It takes energy that would have gone up into the sky and pushes it out toward the horizon.

- *Performance:* **Maximum Range.** This is how you hit repeaters 30 miles away.

- *Use Case:* Desert running, open plains, highway travel.

- *Downside:* It hits every... single... tree branch. *Whack. Whack. Whack.*

Mental Model: The Ground Plane (The Mirror)

This is the concept that confuses everyone. Most mobile antennas are **1/4 Wave** antennas. This means they are physically only one-quarter of the length of the radio wave.

Where is the other three-quarters? **Your car is the other half of the antenna.**

Think of a mirror. If you place a candlestick on a mirror, you see two candlesticks: the real one going up, and the reflection going down. Together, they look like one long object.

Your metal roof is the mirror. The antenna sticks up. The radio wave reflects off the metal roof, creating a "virtual" antenna pointing down. This completes the circuit.

The Implications:

- **Center of Roof:** This is the best spot. The "mirror" is equal in all directions. Your signal goes out 360 degrees perfectly.

- **Hood Mount (Edge):** The "mirror" is behind the antenna (the windshield/roof). Your signal will be strong behind you, but weak in front of you. The pattern is distorted.

- **Plastic Roof (Jeep/Bronco): Problem.** A plastic roof is not a mirror. It is invisible to radio waves. If you mount a standard antenna on a fiberglass roof, you have no Ground Plane. The signal has nowhere to push off of.

  - *The Fix:* You must use a special "No Ground Plane" (NGP) antenna kit, OR mount the antenna on a metal bracket (hood lip, tailgate) that is electrically bonded to the chassis.

Field Tactic: The "Cookie Sheet" Hack If you are setting up a base station at a campsite with a magnetic mount antenna, but you only have a wooden picnic table... Put the antenna on a metal baking sheet (or a pizza pan). Congratulations, you just built a Ground Plane. Your range just doubled.

SAFETY ALERT: The Aluminum Trap If you drive a Ford F-150 (2015+) or a Super Duty (2017+), your truck body is Aluminum. **Magnets do not stick to aluminum.** Do not buy a magnetic mount antenna for these trucks. It will slide right off. You must use a drill-in NMO mount or a dedicated 'fender clamp' bracket.

Action Step: The "Gear Audit"

Before you order anything, we need to talk about **Connector Hell**.

The radio industry cannot agree on a standard plug. If you buy a Baofeng, a Midland, and a dedicated antenna, chances are none of them will fit together without an adapter.

You need to perform a physical audit of your "Male" and "Female" parts.

The "Pin and Hole" Check

Look at the connector on the **Radio Body**.

- **SMA-Male:** Has a center PIN sticking up. (Common on Yeasu, Wouxun).

- **SMA-Female:** Has a center HOLE. (Common on Baofeng, BTech).

Now look at the **Antenna**. It must be the opposite.

Rule of Thumb:

- Buying a **Baofeng**? You usually need an antenna with a **Male** pin (SMA-Female Radio requires SMA-Male Antenna).

- Buying a **Wouxun/Midland/Yaesu**? You usually need an antenna with a **Female** hole (SMA-Male Radio requires SMA-Female Antenna).

The "Gender Changer" Danger You can buy little brass adapters (SMA-M to SMA-F) to bridge the gap.

- **Audit Step:** Order 2 or 3 of these adapters specific to your radio combination.

- **Warning:** Every adapter adds a tiny bit of "insertion loss" (signal weakening) and mechanical stress. Do not stack 4 adapters like a Tower of Babel. If you have to stack adapters, you bought the wrong cable.

The Standard Mobile Connector: PL-259 (UHF)

Most mobile units and roof-mount cables use a much larger connector called the **PL-259** (or SO-239). It looks like a thick silver screw-on plug.

- If you buy a handheld radio and want to connect it to your roof antenna, you will need a **"Pigtail Adapter"**.

- *One End:* Tiny SMA (for the handheld).

- *Other End:* Big SO-239 (for the roof cable).

- *The Cable:* Thin, flexible wire in between to relieve stress.

Your Shopping List:

1. **Primary Radio:** (Mobile 20W+ recommended for drivers, HT for hikers).

2. **The Antenna:** Ghost (Woods) or Whip (Desert).

3. **The Mount:** Lip mount, mag mount, or roof rack bracket.

4. **The Pigtail:** If connecting a handheld to the roof antenna.

5. **USB-C Cables:** One for the car, one for the bag.

Stop looking for the "Perfect Radio." It doesn't exist. Buy the radio that fits your dashboard. Buy the antenna that fits your garage height. Install it cleanly. Then, stop worrying about the gear and start worrying about how to speak on it.

MISSION CONFIG
Hardware Selection & Setup

Cut through the gear paralysis. Whether you are equipping a neighborhood watch or an overland convoy, choose the tool that fits the mission. Don't buy specs; buy reliability.

Handhelds: Burner vs. Operator

The market is flooded. We categorize radios into two tiers. The Burner (Baofeng) is cheap and disposable but has a "Glass Jaw"—it goes deaf near cities due to poor filtering. The Operator (Wouxun/High-End) filters noise and survives the field.

TIER 1: BURNER (Baofeng)
• $25–$40 price point
• Poor noise filtering
• Backup use only

TIER 2: OPERATOR (Wouxun/High-End)
• $100–$250 price point
• Superheterodyne (filters noise)
• IP Rated (Waterproof)

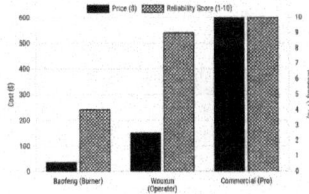

Figure 3.1: Price vs. Signal Reliability

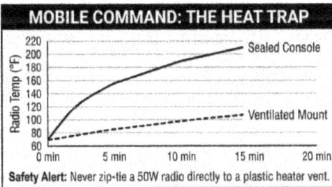

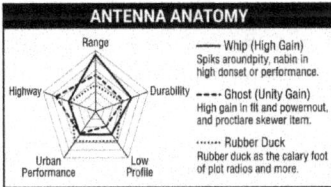

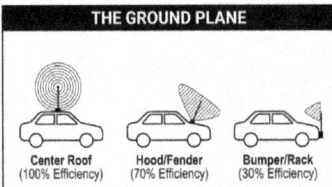

MOBILE COMMAND: THE HEAT TRAP

Safety Alert: Never zip-tie a 50W radio directly to a plastic heater vent.

ANTENNA ANATOMY

Whip (High Gain) Spiks aroundpity, nabin in high donset or performance.
Ghost (Unity Gain) High gain in fit and powemout, and proctlare skewer item.
Rubber Duck Rubber duck as the calary foot of plot radios and more.

THE GROUND PLANE

Center Roof (100% Efficiency)
Hood/Fender (70% Efficiency)
Bumper/Rack (30% Efficiency)

CONNECTOR AUDIT

RADIO BRAND	RADIO CONNECTOR	ANTENNA NEEDED
Baofeng / BTech	SMA-Male (Pin)	SMA-Female (Hole)
Wouxun / Yaesu	SMA-Female (Hole)	SMA-Male (Pin)
Mobile Units	SO-239 (Big Socket)	PL-259 (Big Plug)

The Gender Changer Danger: Every adapter adds signal loss ("Insertion Loss"). Avoid stacking adapters.

Chapter 4

The Visual Command Programming Guide

Conquering CHIRP

OBJECTIVE: OVERCOME THE "PROGRAMMING Gap" to transform the radio from a confusing factory brick into a customized, standardized tool.

Key Deliverable: The "Golden Config"—a standardized memory map cloned across every device in the fleet for seamless operation.

If Chapter 2 (Licensing) was the bureaucratic hurdle, this chapter is the technical firewall.

You have bought the radios. You have your license. You have charged the batteries. Now, you want to put in a specific channel - maybe a local repeater, or a specific frequency your 4x4 club uses.

You open the manual that came with your $35 radio. It is written in broken English, translated from Chinese by a robot that clearly hates you. It tells you to press "MENU + 4 + 7 + UP + ARROW + EXIT" to change one setting. You try it. The radio beeps angrily. You throw the manual across the room.

Welcome to the **Programming Gap**.

Most handheld radios, especially the Baofeng and Wouxun models that dominate the market, are terrible to program by hand. The keypads are too small, the menus are nested like Russian dolls, and the logic is backward. If you have to program a radio in the field using the keypad, you have already failed.

The solution is **CHIRP**.

CHIRP is free, open-source software that allows you to plug your radio into your computer and program it using a spreadsheet. It turns a 4-hour nightmare of button mashing into a 5-minute copy-paste job.

But CHIRP has its own demons. It requires drivers. It requires cables. It requires understanding what "Tone Squelch" actually means.

In this chapter, we are going to build your **"Golden Config."** We will troubleshoot the connection issues that make grown men cry, and we will build a programming file that you can blast into every radio in your fleet, ensuring that when you say "Go to Channel 4," everyone actually goes to Channel 4.

4.1 The Driver Disaster

Before we even open the software, we have to talk about **The Cable**.

When you bought your radio, it might have come with a programming cable in the box. Or maybe you bought a cheap one on eBay for $4.

Throw it away.

I am serious. Walk to your trash can and throw it away.

The "Counterfeit Chip" War

Here is the inside baseball of the radio world. The cable that connects your radio to your computer is not just a wire. Inside the USB plug, there is a tiny microchip that translates the radio's language (Serial/TTL) into your computer's language (USB).

There are two main companies that make these chips:

1. **FTDI (The Good Guys):** Reliable, expensive, works instantly on Windows and Mac.

2. **Prolific (The Complicated Guys):** Older, cheaper, finicky.

Chinese factories, trying to save pennies, flooded the market with **counterfeit Prolific chips**. Microsoft (Windows) got tired of these fake chips causing crashes. So, they updated Windows to automatically detect fake chips and *brick* them - disable them so they don't work.

The Result: You plug in your cheap cable. Windows installs the latest driver. The driver sees the fake chip. The driver kills the connection. You spend 6 hours on Google trying to figure out why your computer won't see the radio.

The Solution: Buy an FTDI Cable

You need to buy a programming cable that explicitly says **"Genuine FTDI Chip"** in the listing. It will cost $20 instead of $5. It is worth every penny.

- **BTech PC03** is the gold standard for Baofeng style (Kenwood 2-Pin) radios.

- **Wouxun** often uses their own high-quality cables.

If you ignore this advice and use the cheap cable, you are entering the "Driver Dance."

Troubleshooting Decision Tree: "My Computer Won't See the Radio"

So, you plugged it in, opened CHIRP, clicked "Download from Radio," and got an error message: *"Radio refused to enter programming mode"* or *"COM Port not found."*

Follow this logic path to fix it.

STEP 1: The "Click" Check

- **The Problem:** The Kenwood 2-pin connector (the plug that goes into the radio) requires a frightening amount of force.

- **The Fix:** Push it harder. Seriously. If you didn't hear a distinct *SNAP*, it's not in. On water-resistant radios (like the Baofeng UV-9G or Wouxun KG-935G), the rubber seal fights you. You have to push until your thumbs hurt.

- **Retest:** Try to download again.

STEP 2: The Device Manager Check (Windows)

- Right-click the **Start Button -> Device Manager**.

- Look for **"Ports (COM & LPT)"**.

- **Scenario A:** You see nothing new when you plug it in.

  - *Diagnosis:* Dead cable or bad USB port. Try a different port.

- **Scenario B:** You see "Prolific USB-to-Serial" with a **Yellow Triangle (!)** icon.

  - *Diagnosis:* The Driver Disaster. Windows updated your driver and killed the fake chip.

  - *The Fix:* You must "Roll Back" the driver.

    i. Right-click the device -> **Update Driver**.

    ii. Select **"Browse my computer for drivers"**.

    iii. Select **"Let me pick from a list..."**.

    iv. Look for an older version (usually from **2008** or **2011**). Select that one.

    v. Click Next. The Yellow Triangle should disappear.

STEP 3: The Mac "Security" Block

- If you are on a Mac, the OS often blocks the driver installation because it's not from the App Store.

- **The Fix:** Go to System Settings -> Privacy & Security. Look for a message saying "System software from developer was blocked." Click **Allow**. Reboot.

STEP 4: The "Volume Knob" Trick

- Some radios need to be turned ON to program. Some need to be OFF.

- **Standard Rule:** The radio must be **ON**, and the volume should be turned up to about 50%.

- **Exception:** Some older radios require you to hold a button while turning them on to enter "Cloning Mode." Check your manual (or Google).

Once you have conquered the cable, CHIRP will open, and you will see a blank spreadsheet. Now, we build the brain.

4.2 The "Golden Config" Spreadsheet

When you successfully "Download from Radio," CHIRP presents you with a grid. It looks like Excel. This is the **Memory** of your radio.

Rows 1 through 22 are usually your standard GMRS channels. Rows above 22 might be blank.

We are going to explain the columns that matter. Ignore the ones that don't (like "Skip" or "Power" for now). The Big Five are: **Frequency**, **Name**, **Tone Mode**, **Tone**, and **Duplex/Offset**.

Column 1: Loc (Location)

This is just the channel number.

- **Tactical Tip:** Make sure your "Fleet" agrees on the numbers. If "Channel 5" is GMRS 5 on your radio, but "Channel 5" is a local repeater on your friend's radio, you are going to have a bad day.

- **Standardization:** Keep 1-22 as the standard GMRS frequencies. Put your special custom channels (Repeaters) in slots 23 through 50.

Column 2: Name (The Alpha Tag)

Cheap radios just show the frequency (e.g., "462.550"). Good radios show a Name (e.g., "GMRS 15").

This column allows you to name the channel. You usually have 6 to 8 characters.

- **Bad Name:** "RPT1" (Vague).

- **Good Name:** "DEN-RPT" (Denver Repeater).

- **Bad Name:** "EMERG" (Vague).

- **Good Name:** "SAR-CH" (Search & Rescue).

Visual Command Rule: Use names that describe the *function* or the *location*, not the frequency.

Column 3: Tone Mode (The Gatekeeper)

This is the single most confusing column in CHIRP. This is where people mess up.

This setting controls the **Privacy Codes** (CTCSS/DCS). Think of the Privacy Code as a **Key**.

- **Setting: "(None)"**

  ○ **Visual:** An Open Door.

  ○ **Tx (Talk):** You shout to everyone.

  ○ **Rx (Listen):** You hear everyone.

  ○ *Use Case:* General comms, channel monitoring, emergency hailing (Channel 20).

- **Setting: "Tone" (Transmit Only)**

  ○ **Visual:** You have a Key to open *their* door, but your door is wide open.

  ○ **Tx:** You send a hidden "hum" (the tone) with your voice. This opens the squelch of the other radio (or repeater).

  ○ **Rx:** You hear *everything* on that frequency, even people not using the tone.

  ○ *Use Case:* **Accessing Repeaters.** Most repeaters require you to send a tone to "wake them up," but they transmit back to you without a tone.

- **Setting: "TSQL" (Tone Squelch - Transmit & Receive)**

  ○ **Visual:** You have a Key, and you locked your own door.

  ○ **Tx:** You send the tone.

  ○ **Rx:** Your radio stays silent *unless* the incoming signal has the matching tone.

  ○ *Use Case:* **Family Convoy.** You want to ignore the construction crew nearby using the same channel. You only want to hear your family.

SAFETY ALERT: If you set your radio to **TSQL** (Tone Squelch), you are filtering out the world. If a stranger is screaming "HELP HELP HELP" on Channel 16 but they aren't using your specific privacy tone, **you will not hear them.**

- *My recommendation:* Run "Tone" (Tx only) or None for most channels. Only use TSQL if you are in a highly congested urban area and the chatter is driving you crazy.

Column 4 & 5: Tone & ToneSql

These columns are where you actually pick the number of the key (e.g., 67.0, 141.3, etc.).

- If **Tone Mode** is "Tone", CHIRP looks at the **Tone** column.

- If **Tone Mode** is "TSQL", CHIRP looks at the **ToneSql** column.

Common Mistake: Setting Tone Mode to "Tone" but changing the number in the "ToneSql" column. The radio will ignore it. You have to fill in the column that matches the Mode.

Column 6: Duplex & Offset (The Repeater Math)

If you are programming a standard channel (Simplex), **Duplex** should be "(None)" or "off".

If you are programming a **Repeater**, you need to tell the radio to "Split" its personality. It listens on one frequency but talks on another.

The GMRS Standard Split:

- **Duplex:** "+" (Positive).

- **Offset:** "5.000" (5 MHz).

What this actually does: Let's say you are on **Channel 15** (462.550 MHz). If you set Duplex to "+" and Offset to "5.0", when you press the PTT button, the radio instantly jumps UP 5 MHz to **467.550 MHz**. This is the **Input Frequency** of the repeater. When you let go of the button, it instantly jumps back DOWN to **462.550** to listen.

Visual Check:

- Look at your CHIRP row.

- Is it a repeater channel (15-22)?

- Does it have a Repeater Tone? -> Set Tone Mode to "Tone" and pick the freq.

- Is Duplex set to "+"?

- Is Offset set to "5.000000"?

- *If yes, you are ready to hit the tower.*

Example: The "Golden Row"

Here is what a properly programmed Repeater channel looks like in CHIRP:

- **Translation:** Channel 23. Named "Denver Repeater". It sends a 100.0 Hz tone to wake up the repeater. It shifts UP (+) by 5.0 MHz when transmitting.

4.3 Creating Your Fleet

You have spent an hour refining your CHIRP spreadsheet. You have named the channels. You have deleted the junk channels (like the Chinese demo frequencies often found in Baofengs). You have added your local NOAA Weather channels (Channels 162.400 etc.) at the bottom of the list with "Duplex: Off".

This file is now your **Master Template**. Save it as Golden_Config_v1.img.

Now we achieve **Fleet Uniformity**.

The "Clone" Process

You want every radio in your group - your handheld, your wife's handheld, your truck mobile unit, your base station - to have the exact same map.

When you scream "Go to Channel Denver!", you don't want your wife asking "Which number is that?" You want it to be Channel 23 on *every single device*.

The Challenge: You cannot always upload a Baofeng file to a Midland radio. The file formats are different. **The Fix:** You use **Copy/Paste**.

1. Open your Golden_Config_v1.img (The Source).

2. Plug in the second radio. Download its (blank/messy) config. This opens a second tab in CHIRP.

3. Go to your Source tab. Highlight all rows (Ctrl+A). Copy (Ctrl+C).

4. Go to the Destination tab. Highlight Row 1. Paste (Ctrl+V).

5. Upload to Radio.

Now, even if the radios are different brands, the logic (Frequencies, Tones, Names) is identical.

The "Scan List" Strategy

One setting you must check before uploading is the **Scan Add** column (sometimes called "Skip").

When you press the "Scan" button on your radio, it cycles through the channels looking for traffic.

- **The Problem:** By default, it scans *everything*. It will scan FRS 1, then FRS 2, then FRS 3...

- **The Annoyance:** Channels 8-14 are low power and usually full of static. Channels 1-7 are full of kids screaming. If you leave these in your scan list, your radio will never shut up.

- **The Fix:** Go to the "Skip" or "Scan Add" column in CHIRP.

  - **Set Channels 1-14 to "S" (Skip).**

  - **Set NOAA Weather Channels to "S" (Skip).** (Otherwise, the scanner will stop on the weather constantly because it is a continuous broadcast).

  - **Set your Tactical Channels and Repeater Channels to " " (Empty/Scan).**

Now, when you hit Scan, your radio becomes a "Hunter." It ignores the garbage and only listens for your convoy or the repeaters you care about.

"Kid-Proofing" (The Keypad Lock)

If you are handing a radio to a child, or even a non-technical adult, they *will* accidentally change the settings. They will grab the radio by the body, and their thumb will mash the "Menu" button, or the "VFO/MR" button which switches the radio from "Channel Mode" (safe) to "Frequency Mode" (dangerous/confusing).

The Software Lock: In CHIRP, go to the **Settings** tab (on the left side). Look for **"Work Mode Settings"** or **"Keypad Lock"**.

- **Auto-Lock:** Set this to "ON". This means if no buttons are pressed for 10 seconds, the keypad locks. You have to hold the # key (usually) to unlock it.

- **Disable VFO:** Some radios allow you to completely disable the "VFO" button via software. Do this. Force the radio to stay in "Channel Mode." This prevents the user from accidentally typing in 440.000 and getting lost in static.

The Physical Lock: If you can't do it in software, teach the "Lock" command. On almost all Baofeng/Wouxun radios, **Press and Hold the # Key** for 2 seconds. A little padlock icon appears on the screen. Do this *before* you hand the radio to anyone.

Workbook Element: The "Visual Command CHIRP Template"

We have done the heavy lifting for you. We have created a standard .csv file that you can import into CHIRP.

This template includes:

- **Channels 1-22:** Standard GMRS (High Power channels set to Wideband).

- **Channels 23-30:** Common "Travel Tone" channels used by overlanders (462.550 with Tone 67.0, etc.).

- **Channels 100+:** NOAA Weather Frequencies (Rx Only).

Scan to Download Printable Copies.

How to Import the Template:

1. Open CHIRP.

2. File -> **Import**.

3. Select the .csv file you downloaded.

4. CHIRP will ask you to map the columns. The defaults are usually correct.

5. Click OK.

Action Step: Stop reading. Go to your computer. Download CHIRP. Order the FTDI cable if you haven't. Do not go into the field with a radio programmed by the factory. The factory doesn't know your mission. You are the Commander. The radio does what *you* tell it to do.

Once you have the fleet programmed, it's time to learn how to actually talk on it. The next chapter covers the "Language of the Field" - how to sound like a pro and clearer comms.

Supplemental: Deep Dive on "Why the Factory Config Sucks"

You might ask, "Why can't I just use the channels that came on the radio?"

If you buy a high-end Midland, you probably can. But if you bought a Baofeng UV-5G or similar:

1. **Narrowband vs. Wideband:** Factory settings often default to Narrowband (NFM) on all channels to be "safe" for FRS rules. GMRS is allowed Wideband (WFM) on the main channels. Wideband sounds louder and richer. If you don't change this in CHIRP, you will sound quiet compared to everyone else.

2. **Random Tones:** Some factory radios come with random privacy tones pre-set on certain channels. You might think you are on Channel 1, but you are actually on Channel 1 with a 67.0Hz tone. You won't hear anyone else who isn't using that tone. You are effectively deaf to the public. **Wipe the slate clean.**

Supplemental: The "Repeaterbook" Integration

A crucial CHIRP feature for the advanced user.

CHIRP has a built-in superpower: **It can talk to "Repeaterbook.com".**

1. In CHIRP, click 'Radio' -> 'Query Source' -> 'Repeaterbook'.

2. Select 'GMRS'.

3. Select your 'State' (e.g., Colorado).

4. It will pull a list of all public repeaters in your state.

5. You can select the ones near you and import them directly into your spreadsheet.

* **Note:** It automatically fills in the Frequency, Offset, and Tone. It saves you from typing errors.

* **Visual Command Rule**: Always verify the data. Sometimes Repeaterbook is outdated.

Listen before you transmit.

Chapter Summary Checklist

Before moving to Chapter 5, you should have:

1. A programming cable with a genuine 'FTDI chip'.

2. 'CHIRP Next' installed and running on your computer.

3. Successfully "Downloaded" the stock config from your radio (and saved it as a backup!).

4. Built your 'Golden Config':

    ○ Standard Channels 1-22 named.

    ○ Local Repeaters added with correct Offsets/Tones.

    ○ NOAA Weather channels added.

    ○ Scan list cleaned up (Skip low power/weather).

5. 'Cloned' this config to at least one other radio.

6. Verified communication between the two radios.

Chapter 5

Privacy Codes & Filters

The "Doorbell" Analogy

OBJECTIVE: SHATTER THE DANGEROUS illusion of "Privacy Codes" while mastering their actual utility as noise filters for convoy sanity.

Key Deliverable: A "Sanity Filter" protocol that blocks interference without compromising operational security awareness (avoiding the "Glass House" trap).

You are at a crowded campground. It's a holiday weekend. Every campsite is occupied by an RV, a tent, and three kids running around with bright yellow walkie-talkies. The airwaves are saturated.

You turn on your radio to Channel 1. *screech* "MOM! MOM! WHERE ARE THE MARSHMALLOWS?" *static* "Back it up... back it up... stop." *beeping* "Marco! Polo!"

It is a cacophony. It is unusable. You can't hear your own family because the noise floor is raised by fifty other users.

So, you look at your manual. It says: **"121 Privacy Codes for Secure Communication."** The box promises that this feature will create a private line just for you. You think, *"Excellent. I'll turn on Privacy Code 12. Now my channel is secure, and nobody can hear us."*

STOP.

You have just fallen for the second biggest lie in the radio industry (right behind the "35-Mile Range").

"Privacy Codes" provide absolutely zero privacy.

Let me repeat that because it is a matter of operational security: **A Privacy Code is not encryption.** It is not a scrambler. It does not hide your voice. It does not prevent others from hearing you.

If you are planning a surprise birthday party, "Privacy Codes" are fine. If you are discussing the location of your hidden supply cache, giving your credit card number to your wife over the air, or discussing a medical emergency involving personal data, "Privacy Codes" will get you compromised.

In this chapter, we are going to strip away the marketing fluff and look at the mechanical reality of **CTCSS** and **DCS**. We are going to replace the word "Privacy" with the word "Filter." By the end of this chapter, you will understand exactly who can hear you, who you can hear, and how to set up a "Family Tone" that actually works without giving you a false sense of security.

5.1 It Is NOT Encryption

To understand why the term "Privacy Code" is a dangerous misnomer, we have to visualize how a radio receiver works. The marketing department wants you to believe you are creating a private tunnel. Physics disagrees.

Mental Model: The Open House vs. The Doorbell

Imagine your radio is a **House**. The **Frequency** (Channel 1) is the **Street Address**.

When you turn your radio to Channel 1 with **NO codes** enabled (Open Squelch):

- **The State:** You have removed the front door, the windows, and the walls. You are living in a gazebo.

- **The Result:** Anyone standing on the street can shout into your living room. You hear everyone. If the wind blows (static), you hear it. If a kid screams three houses down, you hear it.

- **The Reverse:** When you shout, everyone on the street hears you.

Now, let's enable a **"Privacy Code"** (CTCSS/DCS).

What the Marketing says happens: You enter a soundproof bunker deep underground. Nobody can hear you, and you can't hear them.

What ACTUALLY happens: You put the walls and windows back on the house. You close the front door. You install a **Doorbell**.

- **The State:** You are sitting in your living room with the door shut.

- **The Result:** The kid screaming on the street? You don't hear him because your walls muffle the sound. The wind? You don't hear it. The construction crew? Silence. Your speaker remains quiet.

- **The Trigger:** Your wife walks up to the door. She has the specific button (The Tone) to ring your doorbell. *Ding-Dong.* The door opens (Squelch opens). You hear her voice clearly.

- **The Trap:** *While your door is closed, you cannot hear the street. BUT... the windows are made of single-pane glass. Anyone standing on the street can still hear YOU shouting inside.*

The "Glass House" Reality

When you transmit with a Privacy Code, you are shouting out of a house made of glass.

Any radio operator who has their "Door" open (No Codes set) can hear everything you say. This is the default setting for most scanners and many pro-users monitoring the bands.

- **Scanner Guy:** The guy with the police scanner? He hears you.

- **The Rookie:** The kid who just bought a radio and hasn't set any codes? He hears you.

- **The Pro:** The GMRS operator monitoring Channel 16? He hears you.

The **ONLY** thing a Privacy Code does is silence *your* speaker until it hears a matching code. It filters *incoming* audio. It does nothing to *outgoing* audio. It essentially puts earplugs in your ears, but it does not put a muzzle on your mouth.

SAFETY ALERT: The "Stalker" Scenario I have seen this happen in real life. A family is camping. They set their radios to "Channel 1, Privacy Code 5." They think they are on a private line because their radio is quiet. Dad transmits: *"Hey honey, I'm leaving the campsite now, the keys are under the mat, and the kids are alone in the tent."* Dad thinks only Mom heard that because only Mom can talk back to him. In reality, a creep in a van three sites over, scanning Channel 1 with no codes (Carrier Squelch), heard every single word. He knows Dad is gone. He knows where the keys are. He knows the tactical situation.

Visual Command Rule: Never say anything on a radio - encrypted or not - that you wouldn't yell across a crowded parking lot. Even military encryption can be broken. Your $40 Baofeng certainly isn't secure. Treat every transmission as a public broadcast.

So, Why Use Them?

If they don't provide privacy, are they useless? No. They are **Filters**.

They are tools for **Sanity Management**. If you are in a convoy of 4 trucks, and you are driving through Atlanta, the radio waves are full of chatter. Construction crews, kids, interference, and random static bursts. If you don't use a code, your radio will be constantly bursting with static and strangers' voices. You will eventually get annoyed and turn the volume down to save your headache. Then, when your lead truck warns of a tire on the road, you won't hear him because you muted the radio.

By using a code, you silence the chaos. Your radio sits quietly on the dash. It only makes noise when *your* team speaks. That is the utility. It is an **Audio Gatekeeper**, not a security guard. It ensures that when the radio talks, it is someone you actually want to listen to.

5.2 CTCSS vs. DCS

Now that we know *what* they are (Filters), let's look at the two different flavors of filters available on almost every GMRS radio. While they accomplish the same goal, they use different physics to get there.

1. CTCSS (Continuous Tone-Coded Squelch System)

This is the "Old School" analog system. It is often called "PL Tones" (Private Line - a Motorola trademark) or "Sub-Channels."

How it Works: When you hold down the PTT button, your radio transmits your voice. Underneath your voice, deeper than the human ear can hear (sub-audible), it transmits a constant low-frequency hum (a sine wave).

- *Example:* Tone 67.0 Hz.

- Your radio hums a 67.0 Hz note the entire time you are talking.

- The receiving radio listens. If it detects that specific 67.0 Hz hum, it unmutes the speaker. If the hum stops or changes, it mutes.

The Pros:

- **Universal:** Every radio, from a 1980s CB to a modern GMRS unit, supports CTCSS. It is the industry standard.

- **Range:** Because it is an analog hum, it tends to survive static better. If you are on the fringe of reception, the tone usually still gets through even if the voice is scratchy.

The Cons:

- **Slow:** It takes a split second for the receiving radio to detect the hum and open the speaker. This can cut off the first syllable of your sentence. (e.g., instead of "Stop!", they hear "...op!").

- **Limited Keys:** There are only about 38 to 50 standard tones. In a crowded city, chances are high that someone else is using 67.0 Hz, which leads to "False Opens" - where your radio wakes up for someone else's conversation.

2. DCS (Digital-Coded Squelch)

This is the "New School" digital system.

How it Works: Instead of a hum, your radio transmits a stream of binary data (1s and 0s) constantly while you talk. It's like a digital barcode embedded in the audio.

The Pros:

- **More Keys:** There are over 100 standard DCS codes. Less chance of overlap with strangers.

- **Instant:** Digital decoding is often faster/cleaner than analog detection (on good radios).

- **No "Tail":** With CTCSS, when you let go of the button, sometimes there is a brief blast of static (the "Squelch Tail") before the radio closes. DCS usually eliminates this, making the end of the transmission sound like a clean *click*.

The Cons:

- **Fragile (The Digital Cliff):** Digital signals are 'all or nothing.' If you are in a high-static environment (weak signal), the digital data might get corrupted.

- ○ *Analog Fade:* With CTCSS, as you get further away, the audio gets static-filled but remains readable.

- ○ *Digital Drop:* With DCS, as you get further away, the radio simply refuses to open. Or, the audio cuts in and out violently (like a bad cell phone call). This is the "Digital Cliff." You fall off it abruptly.

Visual Command Recommendation: Start with **CTCSS** (Analog Tones). It is more robust for field use where signals vary. Only switch to DCS if you are in a city and experiencing interference on every single CTCSS tone.

The "Rosetta Stone" of Bubble Pack Radios

Here is where the "Tactical Dad" runs into a wall. You have a Baofeng UV-9G (The "Pro" radio). Your brother-in-law shows up with a blister-pack Motorola Talkabout T800 (The "Consumer" radio).

He says: *"I'm on Channel 1, Privacy Code 12."* You look at your Baofeng menu. It asks for a frequency, like 100.0 or D023N. It does not have a "Code 12."

The Problem: Consumer manufacturers (Midland, Motorola, Cobra) decided that "67.0 Hz" was too confusing for the average user. So they created their own arbitrary numbering systems.

- **Motorola Code 1** = 67.0 Hz

- **Midland Code 1** = 67.0 Hz

- **Motorola Code 38** = 250.3 Hz

- **Midland Code 38** = ... might be different.

And then it gets worse. Once they ran out of the 38 Analog tones, they started adding Digital codes to the list to inflate the numbers on the box ("121 Codes!").

- **Code 39** might be the first Digital Code (D023N).

- Or Code 39 might be a non-standard Analog tone.

There is no universal standard for "Code 12." Also, if you want to turn the code **OFF** on a consumer radio, set it to **Code 0**.

Field Tactic: The "Tone Scan" Trick If you cannot figure out what "Code 12" maps to on your pro radio, do not guess. Use the **Tone Scan** feature.

1. Set your Baofeng to the same Channel (e.g., Channel 1).

2. Go to Menu -> R-CTCS (Receive CTCSS).

3. Hit "Scan" (* / Scan button).

4. Have your brother-in-law hold down his PTT button and talk continuously. *"Testing testing 1 2 3..."*

5. Your radio will cycle through the tones until it opens the squelch.

6. It will flash **"100.0"** (or whatever the tone is).

7. **Boom.** You found the translation. Lock that in.

Workbook Element: The Translation Chart We have included a generic translation chart in the Appendix, but *always* verify.

- *General Rule:* Codes 1-38 are usually the standard 38 CTCSS tones in ascending order (67.0 to 250.3).

- *General Rule:* Codes 39+ are usually DCS.

5.3 Tone Squelch vs. Tone Transmit

In Chapter 4, we touched on the CHIRP columns **Tone Mode**, **Tone**, and **ToneSql**. Now we are going to solidify this logic, because getting it wrong means you will be deaf or unheard. This is the single most common configuration error in GMRS.

This is the logic of the **Two-Way Door.**

The Four States of Tone

State 1: Carrier Squelch (CSQ)

- **CHIRP Mode:** (None)

- **Visual:** Both doors are removed.

- **Tx:** You send voice only. (No Key).

- **Rx:** You listen for voice only. (No Lock).

- *Result:* Maximum compatibility. You hear everyone; everyone hears you.

- *Best For:* Hailing frequencies, emergency comms, scanning. If you are lost, be in this mode.

State 2: Tone Transmit (T-TX)

- **CHIRP Mode:** Tone

- **Visual:** You have a Key in your hand, but your door is unlocked.

- **Tx:** You send Voice + Tone. (Using the Key).

- **Rx:** You listen for voice only. (No Lock).

- *Result:* You can open a locked door (like a Repeater), but you can still hear strangers.

- *Best For:* **GMRS Repeaters.** Repeaters almost always require a tone to enter (Input), but do not send a tone back (Output). If you lock your door (State 3) on a repeater channel, you won't hear the repeater responding.

State 3: Tone Squelch (T-SQL)

- **WARNING:** Do not SCAN while in this mode. If you scan with TSQL on, your radio will ignore every conversation that doesn't use your specific key. You will scan past active channels and hear nothing.

- **CHIRP Mode:** TSQL

- **Visual:** You have a Key in your hand, and you have Locked your door with the same key.

- **Tx:** You send Voice + Tone.

- **Rx:** You listen ONLY for Voice + Tone.

- *Result:* The "Private" Line. You only hear radios that have the exact same setting.

- *Best For:* Family camping, convoys in cities, blocking interference.

State 4: Receive Only Tone (Rare)

- **Visual:** You have no key, but your door is locked.

- **Result:** You can't open anyone's door, and nobody can open yours unless they have the key.

- **Utility:** Useless for 99% of users. Avoid this.

Visual Workflow: "Which Mode Do I Choose?"

Ask yourself two questions:

Question 1: Am I talking to a Repeater?

- **YES:** Use **State 2 (Tone Mode: Tone)**.

  - *Why:* Repeaters are usually "Public." They listen for a key to wake up, but they broadcast openly so that any scanner can hear them. If you lock your door (TSQL), you will filter out the repeater and think it is broken.

- **NO:** Go to Question 2.

Question 2: Am I in a "Noise Rich" Environment?

- **YES (City, Event, Crowded Camp):** Use **State 3 (Tone Mode: TSQL)**.

  - *Why:* You need to filter out the garbage.

- **NO (Deep Woods, Desert):** Use **State 1 (Tone Mode: None)**.

  - *Why:* In the middle of nowhere, hearing a stranger is actually useful. It means someone else is nearby (safety/intel). Don't filter out the only other human for 50 miles.

 SAFETY ALERT: The "Double Lock" Fail A common mistake is having a mismatch in the group.

- **Dad's Radio:** Set to TSQL 100.0 (Locked).

- **Kid's Radio:** Set to (None) (Unlocked).

The Result:

- Kid talks -> Dad hears nothing. (Kid didn't use the key).

- Dad talks -> Kid hears Dad. (Kid has no lock).

- *The Green Light of Death:* Dad will look at his radio. He will see the Receive LED light up Green (indicating a signal is hitting the antenna). But the speaker will be dead silent. This is the universal sign of a **Tone Mismatch**. The radio is saying: "I hear a signal, but it doesn't have the secret handshake, so I am not letting it through."

- *Panic ensues.* Dad thinks Kid is lost or the radio speaker is broken. Kid thinks Dad is ignoring him.

- **Visual Command Rule:** Sync the fleet. If one radio is TSQL, ALL radios must be TSQL.

Action Step: The "Family Tone" Setup

You need to pick a standard tone for your family or group. This will be your default setting for your "Tactical Channels" (channels 23-30 in our Golden Config).

1. Avoid the "Default" Tones Humans are lazy. When they buy a Midland radio, they turn it to Channel 1, Code 1.

- **Code 1 (67.0 Hz)** is the most crowded tone in the world. Using this is like trying to have a private conversation in the middle of Times Square.

- **Code 12 (100.0 Hz)** is also very common because it's a nice round number.

2. Pick an "Obscure" Analog Tone Go for the weird numbers in the middle of the list. They are less likely to be used by the family next door.

- *Good Candidates:* 141.3 Hz, 151.4 Hz, 162.2 Hz.

- *Avoid:* 67.0, 88.5, 100.0.

3. Standardize Your "Travel Channel" In your CHIRP config, designate one channel (let's say Channel 25) as **"FAM-TAC"** (Family Tactical).

- **Frequency:** 462.650 (GMRS 19)

- **Tone Mode:** TSQL

- **Tone:** 141.3

Program this into every radio. Label it on your "Comms Card" (from Chapter 2).

The Protocol:

1. **Start on Channel 20 (National Call - No Codes).** Establish contact. *"Unit 1 to Unit 2."*

2. **Move to Secure.** *"Unit 2, switch to FAM-TAC."*

3. **Switch.** Both radios move to Channel 25 (TSQL 141.3).

4. **Verify.** *"Unit 2 on FAM-TAC, how copy?"*

Now you have moved from the "Open Street" to your "Private Room." You have filtered out the noise. You can communicate clearly. But you never, ever forget: **The glass walls are still there.**

Appendix: The "Rosetta Stone" (Abbreviated)

Print this and tape it to your Baofeng battery if you travel with "Bubble Pack" users.

Brand Code	Freq (Hz)	Brand Code	Freq (Hz)	Brand Code	Freq (Hz)
1	67.0	14	107.2	27	167.9
2	71.9	15	110.9	28	173.8
3	74.4	16	114.8	29	179.9
4	77.0	17	118.8	30	186.2
5	79.7	18	123.0	31	192.8
6	82.5	19	127.3	32	203.5
7	85.4	20	131.8	33	210.7
8	88.5	21	136.5	34	218.1
9	91.5	22	141.3	35	225.7
10	94.8	23	146.2	36	233.6
11	97.4	24	151.4	37	241.8
12	100.0	25	156.7	38	250.3
13	103.5	26	162.2	39+	*Usually Digital*

Note: This covers the standard 38 Analog tones used by Motorola/Midland/Cobra/Uniden. Always test to verify.

Troubleshooting: The Green Light of Death If you see your radio's Green LED light up (indicating a signal) but hear silence, you likely have a **Receive Tone (R-CTCS)** set that doesn't match the repeater.

- **The Fix:** Go into your menu and set **R-CTCS to OFF** (or 0.0). This opens your ears to everything the repeater sends.

Chapter 6

Repeater Operations

Extending Your Reach

OBJECTIVE: UNLOCK REGIONAL COMMUNICATION range by mastering the "Split Highway" mechanics of repeaters and the +5MHz offset.

Key Deliverable: The ability to field-program a repeater interface in 60 seconds without a computer, turning a short-range handheld into a 50-mile asset.

You are standing in a deep, pine-choked valley. The air is cold, and the sun dipped below the ridgeline an hour ago. Your friend is in the next valley over, five miles away, waiting for your status update. Between you sits a massive granite ridge, 2,000 feet high and solid rock.

You press the PTT button on your 5-watt handheld radio. *Nothing.* Just the hiss of static. You try standing on a rock. You try holding the radio sideways. *Still nothing.*

The granite eats your signal. UHF waves, as we learned in Chapter 1, are Line-of-Sight. They hit the mountain and stop dead. In this moment, physics has defeated you. You are isolated.

But then, you remember the map. You switch your radio to Channel 23. You press the button and say, *"WRXP-555, radio check."*

Beep-Boop.

A clear, robotic "courtesy tone" chirps back at you. Then, a second later, your friend's voice comes through crystal clear, loud and authoritative, as if he is standing right next to you.

You have just unlocked the superpower of GMRS: The Repeater.

A repeater is a "Force Multiplier." It is a specialized machine that sits on top of that granite ridge, usually powered by solar panels and a bank of heavy batteries. It listens to your weak, 5-watt signal struggling up out of the valley, catches it, cleans it up, amplifies it to a massive 50 watts, and screams it down into the next valley - and often into the next three counties.

Using a repeater turns your $40 handheld radio into a 50-mile communication system. It transforms you from a lone hiker into a node on a regional network.

However, accessing a repeater requires more than just tuning to a channel. It requires understanding the **Split**. Most new users fail here because they treat a repeater like a normal radio. It isn't. It is a machine that demands a specific handshake. If you don't know the secret knock, the machine will ignore you.

In this chapter, we are going to visualize exactly how this machine works, how to find one near you, and how to program it into your radio while standing in the rain without a computer.

6.1 The "Split Highway" Concept

To understand a repeater, you have to stop thinking about "A Channel" and start thinking about "A Highway."

In standard **Simplex** communication (Unit-to-Unit):

- You talk on 462.550 MHz.

- Your friend listens on 462.550 MHz.

- It is a **One-Lane Road**. Only one car can go at a time. If you both try to talk at once, you crash (this is called "doubling"), and nobody hears anything but a garbled mess.

A Repeater is fundamentally different. A Repeater cannot listen and talk on the same frequency at the exact same time. If it did, its own powerful 50-watt transmitter would instantly blow out its own sensitive receiver. It would be like trying to listen to a whisper from across the room while you are screaming through a megaphone. You would deafen yourself.

So, the Repeater uses a **Split Highway**. It separates the traffic into two distinct lanes.

Mental Model: The High Road and The Low Road

Imagine the Repeater is a Giant stationed on a mountain peak. This Giant has excellent hearing and a booming voice.

1. The Listening Ear (Input Frequency) The Giant listens on a specific frequency called the **Input**.

- This is always in the **467 MHz** range (The "High" Road).

- When you talk *to* the repeater, your radio must transmit on this frequency. This is the only frequency the Giant's ears are tuned to.

2. The Shouting Mouth (Output Frequency) The Giant shouts on a different frequency called the **Output**.

- This is always in the **462 MHz** range (The "Low" Road).

- When you listen *to* the repeater, your radio must listen on this frequency. This is where the Giant broadcasts your message to the world.

The Magic Trick: Your GMRS radio handles this switch automatically - *if* and only *if* you program it correctly.

- **Standby State:** When you are just holding the radio, it sits on 462 MHz (The Output), listening for the Giant.

- **Active State:** The *millisecond* you press the PTT button, the radio instantly jumps UP to 467 MHz (The Input) to talk to the Giant.

- **Release State:** When you release the button, it jumps back DOWN to 462 MHz to hear the Giant's reply (or your friend's reply coming through the Giant).

The +5 MHz Offset Rule

The distance between the Input and the Output is exactly **5 MegaHertz**. This separation is critical. It provides enough "spectral distance" so that the Giant's mouth doesn't overpower his ears.

- **Output (Listen):** 462.550 MHz

- **Input (Talk):** 467.550 MHz

- **Difference:** +5.000 MHz

This is called the **Offset**. In the GMRS world, the Offset is *always* **Positive (+)** and it is *always* **5.000 MHz**.

(Note: Ham radio operators have to deal with negative offsets, variable splits, and complex band plans. We have it easy. In GMRS, if you are on a repeater channel, the answer is always +5).

Why This Breaks 90% of New Users Here is the classic scenario of the frustrated Rookie: You look up a repeater. It says "Frequency: 462.550." You type 462.550 into your radio. You press the button. You shout "Radio Check!" **Silence.**

Why? Because you forgot to turn on the Offset.

1. You pressed PTT.

2. Your radio transmitted on 462.550 (The Output).

3. The Repeater is listening on 467.550 (The Input).

4. **Result:** The Repeater didn't hear you. You were shouting at its mouth, not its ear. It's like throwing a ball at the back of the Giant's head. He doesn't know you are there.

5. You hear nothing back. You assume the repeater is broken. It isn't. You just knocked on the wrong door.

6.2 Finding Repeaters

So, where are these magical machines? They are everywhere. There are thousands of GMRS repeaters across the United States, usually located on high ground - mountaintops, water towers, skyscrapers, and even grain silos in the Midwest.

But unlike cell towers, your radio doesn't just "roam" onto them automatically. You have to hunt them down, identify them, and often ask for permission.

The Tool: MyGMRS.com

The absolute gold standard for finding repeaters is **MyGMRS.com**. It is a community-driven database where owners list their machines.

1. Go to the website (or download the app).

2. Create an account (You will need your valid GMRS Call Sign - this keeps the database clean of spammers).

3. Click "Map."

4. Zoom into your area of operations.

You will see pins on the map. Don't just look at the location of the pin; look at the **Coverage Map** circle around it. A repeater on a 10,000-foot peak might cover 80 miles. A repeater on a 100-foot water tower in a valley might only cover 5 miles.

Click on a pin near you. You will see a data sheet like this:

Field	Value	Meaning
Name	Denver 550	The local name (often Refers to the freq).
Output Freq	462.550	The channel you listen to (GMRS 15).
Input Tone	141.3	The "Key" required to open the door.
Output Tone	141.3	The tone the repeater sends back (for your TSQL).
Bandwidth	Wide	Set your radio to WFM.

Understanding "Permission"

This is the social contract of GMRS. Repeaters are privately owned. A regular guy named Dave spent $2,000 of his own money to buy the Motorola Quantar repeater, the hardline coax cable, and the 20-foot fiberglass antenna. He climbed the tower. He pays the monthly electric bill. He pays the site insurance. He is letting you use it out of the kindness of his heart (and for the benefit of community safety). **Respect his rules.**

When you look at a listing on MyGMRS, look for the **Access** status:

1. OPEN System

- **Status:** "Open" or "Public Placed."

- **Rules:** You can use it immediately. The tone is listed right there on the page. Just program it and talk.

- *Etiquette:* Be a good guest. Give your call sign. Don't hog the airtime with a 45-minute discussion about your sciatica. Leave gaps between transmissions so others can break in.

2. PERMISSION REQUIRED

- **Status:** "Permission Required" or "Request Access."

- **Rules:** You cannot just use it. You must click the "Request Access" button on the website.

- The owner (Dave) will get an email. He will check if you have a valid license and if you seem like a normal human being.

- If he approves, he will email you the **Input Tone**.

- *Note:* The listing usually hides the tone until you are approved. **DO NOT** try to "hack" or tone-scan a private repeater. Finding the tone and using it without asking is the fastest way to get banned by the local community. It is the digital equivalent of jumping a fence.

3. CLOSED / PRIVATE

- **Status:** "Private" or "Closed."

- **Rules:** Family and close friends only.

- *Why?* Maybe it's a solar-powered repeater with limited battery capacity. Maybe it's for a specific Neighborhood Watch. If it says Closed, move along.

Field Tactic: The "Travel Tone" Strategy What if you are driving cross-country and don't have time to look up every single repeater in every county? Many "Open" repeaters across the country adhere to a loosely agreed-upon standard called the **Travel Tone** (141.3 Hz). If you are driving through a new city and see a repeater on 462.675 but don't have internet to check the tone, try **141.3 Hz**. It works about 40% of the time on open systems. It's the "Skeleton Key" of the GMRS traveler.

6.3 Programming a Repeater Channel (Field Programming)

You are on a trail. The weather is turning bad. You meet a local wheeler in a Jeep. He rolls down his window and says, *"Hey, heads up, the main road is washed out. We are using the 'Bear Peak' repeater on 675 to coordinate recovery. Tone is 123.0."*

You nod. He drives off. You look at your radio. You don't have your laptop. You don't have CHIRP. You have to program this repeater into your radio **Right Now** using the keypad, while rain drips onto the screen.

This is the "Final Exam" of GMRS proficiency. The exact buttons vary by radio (Baofeng vs. Wouxun vs. Midland), but the **Logic** is universal. You need to map the inputs to the machine.

The "FPP" (Front Panel Programming) Workflow

Step 1: Switch to Frequency Mode (VFO) Most radios have two modes: "Channel Mode" (Memory) and "Frequency Mode" (VFO).

- **Channel Mode:** Shows names like "GMRS 15". It is safe. You can't mess it up.

- **Frequency Mode:** Shows numbers like 462.550. It is the "Wild West." This is where you build new channels.

- *Action:* Press the **VFO/MR** button until you see frequencies.

Step 2: Enter the Output Frequency

- Type 462.675. (This corresponds to GMRS Channel 20).

- *Why:* This is the frequency you will *listen* to.

Step 3: Set the Input Tone (The Key)

- Press **Menu**.

- Look for **T-CTCS** (Transmit Continuous Tone Coded Squelch) or **TX-CODE**.

- *Note:* Do NOT just set R-CTCS (Receive). That locks your ears but leaves your voice unlocked. You need **T-CTCS** to open the repeater's door.

- Select 123.0.

- Press Menu/Confirm.

- *Check:* You should see a little "CT" icon appear on the screen.

Step 4: Set the Offset Direction (SFT-D)

- Look for a menu item called **SFT-D** (Shift Direction).

- Select + (Positive).

- *Note:* Some GMRS-specific radios (like Wouxun KG-935G) hard-code this for you. If you can't find it, the radio might already know to shift up. But on a generic unlocked Baofeng UV-5G, you must set this manually.

Step 5: Set the Offset Amount (OFFSET)

- Look for **OFFSET**.

- Enter 005.000 (5 MHz).

- *Critical Warning:* Many radios display this as 00.000. You want 05.000. If you enter 00.500, you are setting a 0.5 MHz offset (Ham standard) and you will miss the repeater.

Step 6: The "Kerchunk" Test

- Before you save it, test it.

- Press the PTT button for one second. Release.

- **Listen.**

- **The Sound of Success:** You are listening for the **Repeater Tail**. This sounds like a short, sharp blast of static (*pshhht*) or a mechanical *click* that happens about 0.5 seconds *after* you release your PTT. It is the sound of the Giant's transmitter turning off.

- **The Result:**

  - **Tail / Courtesy Tone (Beep):** Success. You opened the machine.

  - **Silence:** You failed. Wrong tone, wrong offset, or out of range.

- **Etiquette Rule:** Do not just click the mic ("Kerchunking"). It is rude. Key up and say, *"WRXP-555 testing."* This identifies you as a respectful operator.

Step 7: Save to Memory

- Now that it works, save it so you don't lose it when you turn the radio off.

- Menu -> **MEM-CH** (Memory Channel).

- Select an empty slot (e.g., Channel 25).

- Press Confirm.

- Switch back to "Channel Mode" (MR) and go to Channel 25.

Visual Command Tip: Most "Bubble Pack" radios (Midland/Motorola) *cannot* be programmed with custom repeater channels in the field via a keypad. They are too simple. They usually have "Repeater Channels 15-22" pre-loaded in a separate bank.

- **Midland MXT Procedure:** You don't program offsets. You just switch to the "RP" channels (15RP through 22RP). The radio already knows the offset. You just go to Menu -> CTCSS and select the tone. It's easier, but less flexible.

Field Tactic: The "Reverse" Check (REV) Most radios have a button labeled **REV** or ***SCAN**. When you hold this button, your radio temporarily swaps the Input and Output frequencies. It makes you listen to the Input (467 MHz).

- **Why do this?** If you press REV and you can hear your friend's voice clearly *without* the repeater, you are close enough to talk Simplex.

- **The Pro Move:** Stop using the repeater. Switch to a Simplex channel (like Ch 16). Save the repeater for when you actually need the range.

Checklist: "Repeater Pre-Check"

Before you try to talk on a repeater, run this 5-second mental audit. If you skip this, you will likely be the guy shouting into the void.

1. Frequency Match: Am I on the right Output Channel? (e.g., 462.xxx).

2. Tone Match: Do I have the correct **Transmit (Input)** tone set? (Double check T-CTCS vs R-CTCS).

3. Offset Active: Do I see a little "+" icon on my screen? (If not, I am talking Simplex and the repeater won't hear me. I am talking to its mouth).

4. Bandwidth: Is my radio set to **Wide (WFM)**?

   ○ *Why:* Most repeaters are Wideband. If your radio is set to Narrow (NFM), your audio will sound incredibly quiet, like you are whispering from inside a box. The repeater's "gate" might even cut you off because your audio isn't loud enough to keep it open.

5. Listen First: Is anyone using the repeater right now? Don't interrupt a conversation just to say "Radio Check." Wait for a break.

The Golden Rule of Repeaters: A Repeater is a shared resource. It is a "Community Campfire." When you step up to the campfire, don't start screaming. Introduce yourself. *"WRXP-555, monitoring."* This simple phrase tells the locals you are there, you are licensed, and you are listening. It invites conversation without demanding it. It marks you as a "Citizen Operator," not a "Lid" (a bad operator).

Troubleshooting: The Green Light of Death If you see your radio's Green LED light up (indicating a signal) but hear silence, you likely have a **Receive Tone (R-CTCS)** set that doesn't match the repeater.

- **The Fix:** Go into your menu and set **R-CTCS to OFF** (or 0.0). This opens your ears to everything the repeater sends.

Chapter 7

Protocol & Etiquette

Sounding Like an Operator

OBJECTIVE: TRANSFORM FROM A mic-shy purchaser into a confident communicator using professional, concise logic.

Key Deliverable: Three "Glovebox Scripts" for instant reference.

7.1 The Anti-Sad Ham Manifesto

Let's get one thing straight before you key that microphone: You are not a trucker from 1978, and you are not playing a character in a bad Hollywood war movie. You are a parent securing your family, or a convoy leader guiding vehicles over an obstacle.

The biggest hurdle to effective radio use isn't "Line of Sight" physics or CTCSS privacy tones - it is **Mic Fright**. This is the paralyzing fear that you're going to say the wrong thing and get yelled at by the "Radio Police."

In the amateur radio world, there is a subset of hobbyists often derisively called "Sad Hams." These are the gatekeepers. They are the guys who sit in their basements with $5,000 worth of equipment, listening to the silence, waiting for someone to make a minor procedural error so they can lecture them. They treat the airwaves like a country club where you have to wear a tie to get in.

We are not them.

In the *Visual Command System*, a radio is a tool, not a lifestyle. It sits in the hierarchy of survival gear right next to your fire extinguisher and your winch. You don't have a secret language for using a fire extinguisher, and you don't need one for GMRS.

The Problem with "10-Codes"

You have likely heard "10-4" (Message Received) or "10-20" (Location). These are called 10-Codes. They were popularized by law enforcement and CB radio users decades ago to shorten transmission times on crowded channels.

Do not use them.

Why? Because they are dangerous. 10-Codes vary by jurisdiction. In one county, a "10-50" might mean a car crash; in the next county over, it might mean an officer needs a bathroom break. When you are dragging a trailer down a scree slope or trying to locate a lost child in a state park, ambiguity is the enemy.

If you say, "I've got a 10-33 here," and your wife thinks you mean "Emergency Traffic" but you actually meant "Spotter Required" based on a list you saw on the internet, you have introduced friction into a safety system. Friction gets people hurt.

The "Breaker Breaker" Trap

If you start a transmission with "Breaker Breaker, one-nine," you have immediately signaled to everyone within five miles that you have no idea what you are doing. This is Citizens Band (CB) slang from a bygone era. It has no place on GMRS. It's the verbal equivalent of wearing your shoes on the wrong feet.

The Solution: Plain Language

The United States government, specifically the Department of Homeland Security and FEMA, officially moved away from 10-Codes and jargon in 2006 (under the NIMS/ICS protocols). They mandated **Plain Language**.

If you want to sound like a Tier-1 operator, don't use code. Speak English.

- **Instead of:** "Can I get a 10-20 on the Lead Vehicle?"

- **Say:** "Lead Vehicle, what is your location?"

- **Instead of:** "That's a big 10-4, good buddy."

- **Say:** "Copy that." or "Received."

- **Instead of:** "Breaker One-Nine for a Radio Check."

- **Say:** "Radio Check."

Plain Language is the ultimate "anti-elitist" protocol. It works for your 6-year-old child, your spouse, and the Search and Rescue team that might be monitoring the channel. It strips away the ego and leaves only the information.

The Visual Command Rule of Brevity

While we use plain language, we do not use *excessive* language. A radio channel is a "single-file hallway." Only one person can walk through (talk) at a time. If you stand in the hallway having a casual conversation, you are blocking the flow for everyone else.

Think before you press the button (PTT). Visualize your sentence like a text message. If it would cost you $5.00 to send the message, how would you edit it down?

The "Sad Ham" Way (WRONG)	The Visual Command Way (RIGHT)
"Uh, yeah, this is WXYZ123, just calling out to the unit in the red Toyota, wondering if you guys saw that turn off back there, over."	"Red Toyota, you missed the turn. Turn around."
"Breaker, breaker, can I get a radio check? I just bought this new Baofeng and want to see if I'm hitting the repeater."	"Radio Check."
"10-4, I will proceed to your 10-20 effectively immediately."	"Moving to you now."

The Takeaway: Sounding like a pro doesn't mean sounding complicated. It means sounding *clear*.

7.2 Scripts for Anxiety-Free Comms

When adrenaline spikes - like when a vehicle is sliding sideways or a kid wanders off - your cognitive function drops. You don't rise to the occasion; you fall to the level of your training.

If you don't have a script, you will stutter, mumble, or freeze.

We use the **"Hey You, It's Me"** protocol. This is the global standard for professional communications (aviation, military, marine).

1. **Hey You:** Address the person you want.

2. **It's Me:** Identify yourself.

3. **The Message:** The instruction or question.

Below are the three "Glovebox Scripts." Print these out. Tape them to the dash. Drill them.

Script A: Requesting a Radio Check

You've just turned your radios on. You need to verify they are working before you lose cell signal. Do not tap the microphone (this damages the element) and do not blow into it.

The Logic: You are throwing a ball into the dark. You are waiting for someone to throw it back.

The Script:

YOU: "Radio Check." (Wait 3 seconds). **YOU:** "Any station, Radio Check, this is [Your Call Sign]."

If someone replies:

THEM: "Loud and clear, [Call Sign]." **YOU:** "Thanks, [Your Call Sign] clear."

Note on Etiquette: If you hear someone ask for a radio check, answer them! A simple "I hear you, this is [Your Call Sign]" is all that is required. It is a courtesy that confirms their safety equipment is functional.

Script B: Calling a Specific Person

You are the "Chase Vehicle" and you need to talk to "Lead Vehicle" (Dad).

The Logic: Open the channel, establish the link, *then* send traffic. Do not dump the information in the first transmission, because if they weren't listening, they missed it.

The Script:

YOU: "Lead Vehicle, from Chase. Over."

(Wait for them to wake up and grab the mic)

THEM: "Go ahead, Chase."

YOU: "There is a large rock in the left track, suggest swinging wide right. Over."

THEM: "Copy. Wide right. Out."

Why this works:

- **"Lead Vehicle, from Chase":** This alerts *everyone* on the channel that the conversation is for Lead, and it is coming from Chase. Everyone else knows to stay quiet.

- **"Over":** This is a "handshake." It hands the frequency to the other person.

- **Read-back:** Notice "Lead" repeated the instruction ("Wide right"). This confirms they understood the specific safety instruction. This is called "Closed Loop Communication."

Script C: Emergency Break-In

This is the "Nuclear Option." Use this only when life, limb, or property is at immediate risk, and the channel is currently busy with other people talking.

Perhaps a group of hikers is chatting on the channel you monitor, but your truck just caught fire. You need to interrupt them.

The Logic: You must shatter the current conversation rhythm to gain attention.

The Script:

YOU: "BREAK, BREAK, BREAK. Emergency Traffic."

(The channel should go silent instantly).

YOU: "This is [Your Call Sign], I have a medical emergency at [Location]. Is anyone copying? Over."

Rules of the "Break":

1. **Never** use "Break" for non-emergencies (like asking for a mic check). It is like pulling a fire alarm to see if it works.

2. If you hear "Break, Break," stop talking immediately. Listen. If you are the strongest station, answer them and offer assistance.

The "Over" vs. "Out" Distinction

In movies, you hear "Over and Out." In reality, those are contradictory terms.

- **Over:** "I am done talking, but I expect a reply." (Handing the ball to you).

- **Out:** "I am done talking, and the conversation is finished." (Putting the ball in the box).

- **Clear:** "I am leaving this channel."

Correct: "Copy that, see you there. Out." **Incorrect:** "Copy that, Over and Out." (This sounds like "I'm expecting a reply, but I'm hanging up.")

7.3 The NATO Phonetic Alphabet

GMRS operates on UHF frequencies. While FM is clear, static happens. Obstacles happen. Wind noise happens.

The English language is terribly designed for radio.

- "B" (Bravo)

- "C" (Charlie)

- "D" (Delta)

- "E" (Echo)

- "G" (Golf)

- "P" (Papa)

- "T" (Tango)

- "V" (Victor)

- "Z" (Zulu)

Nine letters in the alphabet end in the "EE" sound. If you are spotting your wife over an obstacle and shout "Go feet!" she might hear "Go Deep," "Go P," or "Go V."

If you say "My callsign is WRKC290," the "C" might sound like "Z" or "V" or "E".

To solve this, we use the **NATO Phonetic Alphabet**. This is not militaristic cosplay; it is audio engineering designed to ensure that no two words sound alike, even in heavy static.

The Visual Command Cheat Sheet

Do not try to memorize this overnight. Print this page. Cut out this table. Tape it to the back of your handheld radio or your dashboard.

Letter	Word	Pronunciation	Why? (Audio Distinction)
A	Alpha	AL-fah	Hard 'L' sound distinguishes from others.
B	Bravo	BRAH-voh	Strong 'BR' start, distinct 'O' finish.
C	Charlie	CHAR-lee	'CH' is a hard consonant sound.
D	Delta	DEL-tah	Two hard syllables. Impossible to confuse.
E	Echo	ECK-oh	Short, sharp start.
F	Foxtrot	FOKS-trot	Two distinct words combined.
G	Golf	GOLF	Hard 'G', ends in 'F'.
H	Hotel	hoh-TEL	Distinct accent on second syllable.
I	India	IN-dee-ah	Three syllables, unmistakable.
J	Juliet	JEW-lee-ett	Soft 'J', sharp 'T' end.
K	Kilo	KEY-loh	Hard 'K'.
L	Lima	LEE-mah	Liquid 'L' sound.
M	Mike	MIKE	One syllable, sharp 'K' end.

Letter	Word	Pronunciation	Why? (Audio Distinction)
N	November	no-VEM-ber	Three syllables, unlike 'Mike'.
O	Oscar	OSS-cah	Distinct 'O' start.
P	Papa	PAH-pah	Plosive 'P' sounds cut through noise.
Q	Quebec	keh-BECK	Hard 'Q' and 'K' sounds.
R	Romeo	ROW-me-oh	Rolling flow.
S	Sierra	see-AIR-rah	Sibilant 'S' sound.
T	Tango	TANG-go	Hard 'T' and 'G'.
U	Uniform	YOU-nee-form	Distinct 'U' start.
V	Victor	VIK-tah	Sharp 'K' in the middle.
W	Whiskey	WISS-key	Distinct 'W' and 'K'.
X	X-ray	ECKS-ray	Only one starting with 'E' sound besides Echo.
Y	Yankee	YANG-key	Twangy sound cuts static.
Z	Zulu	ZOO-loo	Deep vowels.

How to Practice Without Looking Weird

You don't need to speak in full phonetic sentences ("I AM GOING TO THE GOLF-RRR-OSC-EEE-RRR-Y STORE"). That is strictly "Sad Ham" territory.

You use phonetics **only for critical data**:

1. **Call Signs:** "This is W-R-K-C-2-9-0" becomes "Whiskey Romeo Kilo Charlie Two Nine Zero."

2. **Map Coordinates:** "Grid Square Alpha Four."

3. **Spelling clarifications:** "Meet at the trail head. That is trail B as in Bravo."

The **"Say Again" Rule:** If you don't understand a transmission, do not say "What?" or "Huh?". Say: **"Say Again."** If you need them to use phonetics because the signal is bad, say: **"Say Again, verify with phonetics."**

Action Step: The "Driveway Drill"

Reading this chapter does not make you a competent operator. Pushing the button makes you a competent operator.

Anxiety comes from the unknown. If the first time you try to use "Script B" is when you are stuck in a mud hole with your wife yelling at you and the rain pounding on the roof, you will fail. You will revert to shouting "CAN YOU HEAR ME?" and getting frustrated.

We are going to kill that anxiety today, in the comfort of your driveway.

The Setup

1. **Participants:** You (The Tactical Dad) and your Partner (The skeptical spouse or eager child).

2. **Gear:** Two handheld radios (HTs).

3. **Location:** Your driveway and inside your house (or down the street).

The Drill Execution

Step 1: The Separation Give one radio to your partner. Send them inside the house or to the end of the block. You stay at the vehicle. *Ensure both radios are on the same channel.*

Step 2: The Radio Check (Script A)

- **You:** "Radio Check."

- **Partner:** "Loud and clear. This is [Partner Name]."

- **You:** "Received. Out."

Why: This confirms battery, frequency, and volume.

Step 3: The Specific Call (Script B)

- **You:** "Base, this is Mobile 1. Over."

- **Partner:** "Go ahead, Mobile 1."

- **You:** "I am initiating the convoy test. Requesting current time. Over."

- **Partner:** "Time is [Current Time]. Over."

- **You:** "Copy [Current Time]. Out."

Why: This practices the "Hey You, It's Me" handshake. It feels awkward at first. Do it anyway. It builds the muscle memory of *waiting* for the reply before dumping data.

Step 4: The Phonetic Spelling

- **You:** "Base, Mobile 1. I need you to record a license plate number. Ready to copy?"

- **Partner:** "Go ahead."

- **You:** "License plate is: Alpha, Bravo, Seven, Niner, Zulu, Tango. Over."

- **Partner:** "I copy Alpha Bravo Seven Nine Zulu Tango?"

- **You:** "Correct. Out."

Why: This forces you to glance at the cheat sheet and use the words "Alpha" and "Zulu" instead of "Apple" and "Zebra."

The "Hot Mic" Awareness Check

Before you finish the drill, teach your family about the "Hot Mic." Have your partner hold the PTT button down and try to listen to you. They won't hear anything. **Rule:** *When you are talking, you are deaf.* GMRS is "Simplex" (one way at a time). If you hold the button down to think ("Uhhhhh...."), you are jamming the frequency and nobody can tell you to stop. **Teach:** Push. Pause (1 second). Talk. Release immediately.

Debrief

Once the drill is done, put the radios back in the chargers. You have now established a baseline of competence. The next time you hand that radio to your spouse on a trail, they won't look at it like it's an alien artifact. They will remember the "Driveway Drill." They will know the flow.

You have moved from "Person with a Radio" to "Operator."

Chapter 8

The Family PACE Plan

SOPs for Safety

OBJECTIVE: REPLACE "PANIC AND guessing" with a pre-loaded decision algorithm.

Key Deliverable: A fill-in-the-blank PACE Card for your visor.

8.1 PACE Defined: The Algorithm of Survival

You are five miles down a washboard trail. The sun is setting, casting long, confusing shadows across the desert floor. You pick up the radio to tell your spouse (driving the chase vehicle) that you are taking the left fork toward the campsite. You press the button. Silence. You press it again. Static.

What do you do?

Do you stop in the middle of the trail? Do you turn around and risk a multi-point turn on a narrow ledge? Do you wait? Does she know to wait? Or does she take the right fork because "it looked smoother," separating your convoy by two ridges and ten miles of darkness?

In that moment of silence, your heart rate spikes. Your IQ drops. This is not the time to invent a plan.

Without a plan, you are gambling. In the *Visual Command System*, we do not gamble with family safety. We use **PACE**.

PACE is an acronym borrowed from Special Operations, used by Green Berets and Forward Air Controllers. But you don't need a beard and a suppressed rifle to use it. It is simply a logic tree for communication failure. It answers the single most important question in the backcountry: *"What exactly do we do when this thing stops working?"*

The Hierarchy of Comms

Most people have a Plan A. Smart people have a Plan B. A PACE plan gives you four layers of redundancy, sorted by friction and reliability. It is a contract between everyone in your group.

P: Primary (The "Easy Button")

This is your default method of communication. It is the path of least resistance.

- **In the Suburbs:** Your Primary is likely your **Cell Phone**. It's easy, high-bandwidth, and you already have it. However, cell towers are fair-weather friends. In a disaster or a valley, they are the first to fail.

- **In the Convoy:** Your Primary is your **GMRS Radio on a specific simplex channel** (e.g., Channel 16). It provides instant, one-to-many communication without dialing. It creates a "party line" where everyone hears everything, providing total situational awareness.

- *Visual Command Analogy:* The Primary is the paved highway. It's fast, smooth, and comfortable. But if there's a roadblock or a washout, you cannot just sit there. You need an exit ramp.

A: Alternate (The "Detour")

The Alternate is a backup system that provides similar capability but on a different path. It requires a deliberate, pre-briefed action to switch.

- **Scenario:** Channel 16 is suddenly flooded with a construction crew using high-power commercial radios, or a group of kids playing hide-and-seek. Or perhaps you accidentally enabled a "Privacy Tone" that your partner doesn't have, locking them out.

- **The Fix:** You switch to **GMRS Channel 20 (Simplex)**.

- **The Key:** Everyone must know *before* you leave the driveway that Channel 20 is the Alternate. If you switch and your spouse doesn't, you are talking to yourself. This isn't just about changing numbers; it's about having the discipline to make the switch in unison.

- *Visual Command Analogy:* The Alternate is the dirt service road running parallel to the highway. It's a bit bumpier (you have to stop and change channels), but it gets you to the same destination efficiently.

C: Contingency (The "Force Multiplier")

The Contingency is used when the environment has defeated your Primary and Alternate. Usually, this means terrain has blocked your Line of Sight. Simplex radios cannot punch through a granite mountain.

- **Scenario:** You drove over a ridge. Your spouse is still in the valley. Your simplex signal is hitting the dirt and dying. You cannot hear them, and they cannot hear you.

- **The Fix:** You switch to the **Local Repeater** (e.g., The "Saddleback 700" Repeater).

- **Why it works:** The repeater is high up on a tower or peak. You talk *up* to it; it blasts your voice *down* to the valley, bridging the gap that simplex couldn't cross. This requires foresight - you must have the repeater frequencies and input tones programmed before you lose service.

- *Visual Command Analogy:* This is the helicopter. When the road is washed out and the service road is flooded, you fly over the obstacle.

E: Emergency (The "Nuclear Option")

The Emergency method is for when the world has truly gone wrong. Radios are dead, broken, or batteries are flat. Or perhaps someone is injured and you need outside help immediately. This is the layer of last resort.

- **Scenario:** Total electronic failure, vehicle rollover, or medical critical.

- **The Fix: Satellite Messenger** (Garmin inReach/Zoleo) or **Physical Rendezvous** (Walking back to the last known intersection).

- **The Mental Anchor:** Establishing a physical "Meet Point" (e.g., "If we lose all comms, we meet back at the trailhead") provides a massive psychological advantage. It stops people from wandering aimlessly.

- *Visual Command Analogy:* This is the flare gun or the hiking boots. It is slow, expensive, or physically demanding, but it works when nothing else does.

The "If-This-Then-That" Logic Tree

The beauty of PACE is that it removes emotion. You don't get frustrated when Channel 16 fails; you simply execute the logic. You don't scream at the radio; you switch the channel.

The Convoy Protocol:

"Lead vehicle to convoy. Primary is compromised. Executing PACE. Moving to Alternate (Channel 20). Acknowledge."

If you get no response on Primary, you blindly switch to Alternate and try again. If the plan was briefed correctly, your team will be there waiting for you.

8.2 Kid Protocols: Managing the "Junior Operators"

Handing a GMRS radio to a child is a rite of passage. It gives them autonomy, a sense of responsibility, and a tether to safety. However, an untrained child with a radio is a liability - and potentially a beacon for trouble.

We need to establish the "Junior Operator" Rules of Engagement to ensure the radio remains a tool, not a toy.

Rule 1: The "Whistle Rule" (Analog Backup)

Batteries die. Knobs get bumped. Kids drop electronics in creeks. If your child's safety relies 100% on a $30 piece of plastic and a lithium-ion battery, you have failed the PACE plan.

The SOP: Every child issued a radio is also issued a high-decibel safety whistle (pealess, like a Fox 40). Lanyarded to their neck or clipped to their pack (never loose in a pocket).

- **The Rule:** "If you push the button and Daddy doesn't answer three times, stop pushing. Blow the whistle three times. Stay put."

- **Why:** A whistle cuts through wind, river noise, and forest canopy differently than a radio voice. It requires no electricity and no Line of Sight. It is the 'Emergency' layer of their personal PACE plan.

- **The Universal Signal:** Teach them that three blasts means "SOS" or "I need help." Two blasts means "Come here." One blast means "Where are you?"

Rule 2: "Creepy Chatter" Protocol

GMRS is a public band. You are sharing the airwaves with everyone within 5 to 50 miles. This includes helpful grandpas, bored truckers, and, unfortunately, creeps.

If your child has a "cute" voice, they may attract attention. A stranger might jump on the channel and say, *"Hey there, little ranger, where are you camping?"* or *"What's your name? I have a dog here."*

This is not paranoia; it is operational security (OPSEC). A predator uses information to close distance. Don't give it to them.

The Training: Teach your children to identify "Stranger Traffic."

1. **The Trigger:** Anyone asking for their name, location, or who they are with (other than Mom/Dad).

2. **The Action: Do Not Reply.** Silence is the best armor. A stranger cannot find you if they cannot hear you.

3. **The Notification:** The child immediately brings the radio to you, or switches to the designated "Safe Channel" (Alternate) if they are old enough to manage it.

The "Code Word" Concept: Give your child a code word that means "Switch channels immediately."

- **Dad says:** "Jurassic Park."

- **Kid does:** Clicks radio knob one click to the right (to the pre-set Alternate channel).

- **Why:** This allows you to move the conversation away from the creeper without engaging the stranger or alerting them that you are moving. It maintains your control over the situation.

Rule 3: Radio Discipline for Kids (The "No DJ" Rule)

Kids love to hear their own voice. They love to push the "Call" button that makes the ringing sound. They love to key the mic and breathe, or narrate their hike.

- **The Consequence:** If they jam the frequency with nonsense, you cannot hear the warning about the bear on the trail, the approaching Jeep, or the weather alert.

- **The Fix:** "The radio is a seatbelt. We don't play with seatbelts. We wear them for safety. If you play DJ, the radio goes in Daddy's pack, and you have to hold hands the rest of the hike."

- **The Mechanical Fix:** Enable the 'Keypad Lock' before handing the radio to a child. This prevents them from accidentally changing the channel while hiking or playing.

- **The "Listen First" Habit:** Teach them to listen for 3 seconds before they talk. If someone else is talking, they must wait. This prevents "doubling," where two signals cancel each other out, resulting in a horrible screeching noise.

8.3 The "Lost Comms" Procedure

The most terrifying sound in the backcountry is absolute silence when there shouldn't be. You are waiting for your scout to check in. You call. Nothing. You call again. Nothing.

The human brain immediately jumps to catastrophe. *"They fell off a cliff." "They are hurt."* Usually, they just drove behind a rock, their battery died, or they bumped the volume knob.

To prevent panic-induced errors (like driving off a trail to search or splitting up the group), we use a **Visual Timeline** for Lost Comms.

The Timeline of Silence

T-Minus 0:00 to 0:01 (The "Idiot Check")

Before you panic, check your own gear. 90% of radio failures are user error.

- Is *your* **radio on?**

- Is *your* **volume up?** (Did you bump it getting out of the truck? Turn it all the way up until you hear the beep/static).

- **Are you on the correct channel?** (Did you accidentally sit on the button and switch to Channel 4?)

- **The Keypad Lock Betrayal:** Did you accidentally lock the keypad? Or worse, is the keypad *unlocked* and you bumped a setting? Many radios have a 'Lock' icon. Check it.

- **Action:** Turn radio off and on. Check channel. Check volume. Unlock/Lock keypad.

T-Minus 0:01 to 0:05 (The "Static Check")

If your gear is good, assume the environment is the problem. Your vehicle is a cage of metal and electrical noise.

- **Action:** Stop the vehicle or stop walking.

- **Action: Kill the engine and accessories.** Alternators, spark plugs, and **cheap LED light bars** generate RF interference that can drown out weak signals. You might not hear the noise, but your radio does.

- **Action:** Turn your squelch *down* to zero (or press the 'Monitor/Moni' button) to listen to raw static. A weak signal might be breaking the squelch threshold but not opening it. You are looking for a faint voice in the white noise.

- **Call:** "Convoy, this is Lead. Radio check. Over."

T-Minus 0:05 to 0:15 (The "Seek Higher Ground")

Radio waves are light waves. They cannot go through mountains, dense wet foliage, or rock walls. If you have lost comms, you likely lost Line of Sight. You are in a "Radio Shadow."

- **Action:** Do not drive deeper into a valley or canyon. Drive *up*. Get to a ridgeline or a clearing. Even 10 feet of elevation can restore a signal. Standing on the roof of your truck adds 6 feet to your antenna height - that can double your horizon.

- **Protocol:** "Convoy, Lead is moving to high ground for comms. Hold position." (Say this even if you think they can't hear you - they might be receiving but unable to transmit).

T-Minus 0:15+ (The "Blind Transmit" & Rendezvous)

If 15 minutes pass with no contact, you are no longer in a "glitch." You are in a "situation."

- **Action:** Execute the PACE plan logic.

  - Try Alternate Channel.

  - Try Contingency Repeater.

- **The Blind Call:** Transmit your intentions *in the blind*. You assume they can hear you, even if you can't hear them.

"Chase, this is Lead. Lost comms. I am holding at the trail intersection marked 'Scenic Overlook' for 30 minutes. If no contact, I will backtrack to the trailhead. Out."

- **Why:** They might hear you perfectly but have a broken microphone (or a shy child holding the radio). Telling them exactly where you are and what you are doing prevents them from driving in circles looking for you. It gives *them* a plan, even if they can't talk back.

Workbook Element: The PACE Card

You will not remember the repeater frequency when your truck is overheating, the sun is down, and your kids are crying. Write it down.

Instructions:

1. Print this card (or copy it onto an index card).

2. Fill it out with a Sharpie.

3. **Weatherproof it:** Cover the card in clear packing tape to laminate it. A soggy card is useless.

4. Tape it to your dashboard or visor.

5. Make a copy for every radio user in your group.

```
+---------------------------------------------------------------+
|                VISUAL COMMAND :: PACE CARD                    |
+---------------------------------------------------------------+
| MISSION/TRIP: _____ |
| GROUP CALL SIGN: _____  |
+---------------------------------------------------------------+
|  P  | PRIMARY (The Main Channel)                              |
|     |   Channel: _____  (Name: _____)                  |
|     |   Tone: _____                                       |
+---------------------------------------------------------------+
|  A  | ALTERNATE (The Backup / Simplex)                        |
|     |   Channel: _____  (Name: _____)                  |
|     |   Tone: _____                                       |
+---------------------------------------------------------------+
|  C  | CONTINGENCY (The Repeater / High Ground)                |
|     |   Channel: _____  (Name: _____)                  |
|     |   Rptr Input Tone: _____                            |
+---------------------------------------------------------------+
|  E  | EMERGENCY (Total Failure)                               |
|     |   [ ] Sat Messenger   [ ] Cell Phone (911)              |
|     |   [ ] Physical Meet Point: _____  |
+---------------------------------------------------------------+
| LOST COMMS PROTOCOL:                                          |
| 1. Check Vol/Ch/Lock.                                         |
| 2. Kill Engine / Check Static.                                |
| 3. Move High.                                                 |
| 4. Wait 15 Mins.                                              |
| 5. Return to: _____ |
+---------------------------------------------------------------+
```

The "Briefing" Habit

Before you put the car in 'Drive', hold up this card. Point to it. Ask your passenger: *"What is our Alpha channel today?"* If they don't know, tell them.

This takes 30 seconds. It saves 3 hours of panic. You are now operating with a system, not just a gadget.

Chapter 9
The Mobile Build
Overland Install Guide

OBJECTIVE: UPGRADE FROM A handheld "toy" to a 50-watt "Command Center" without burning down your truck.

Key Deliverable: A vehicle that communicates as well as it crawls.

Introduction: The Difference Between a Radio and a System

If you have followed the previous chapters, you likely have a handheld radio (HT) clipped to your sun visor or rattling around in a cup holder. That HT is a great tool. It's your "scout" weapon - perfect for spotting a driver over an obstacle or hiking up a ridge. But if you are serious about the *Visual Command System* - if you are the Convoy Lead, the Chase Vehicle, or the primary safety node for your family - the handheld is not enough.

You need a Mobile Unit.

A Mobile Unit (or "Base Station") is a high-power (20 to 50 watts) radio hard-mounted into your vehicle, connected to the external antenna we mounted in Chapter 6, and powered directly by your vehicle's electrical system. This isn't just an accessory; it's infrastructure.

This upgrade changes the game physically and tactically.

- **Wattage:** You go from 5 watts (throwing a baseball) to 50 watts (firing a cannon). This punches through foliage and bounces signals off canyon walls with authority.

- **Reliability:** No more dead batteries in the middle of a trail. Your power source is now a 60-pound lead-acid battery constantly recharged by the engine.

- **Audio Authority:** A loud, external speaker that cuts through the noise of mud-terrain tires, diesel engines, and screaming kids. You will hear every call, even with the windows down.

- **The Faraday Cage Effect:** Your vehicle is a metal box. Using a handheld *inside* the car is like trying to shout through a steel bucket. A mobile unit moves the "mouth" (antenna) outside the bucket, instantly doubling your effective range before you even add power.

However, installing a mobile radio is not like plugging in a USB charger or wiring a dashcam. You are integrating a high-power Radio Frequency (RF) transmitter into a complex web of automotive electronics. Do it wrong, and you'll have a radio that whines like a dying cat every time you accelerate. Do it really wrong, and you'll burn out the transmitter, melt your wiring harness, or drain your starter battery in the backcountry.

This chapter is the mechanic's manual for the Tactical Dad. Put on your gloves. We are going under the hood.

9.1 Power & Noise: The Lifeblood of Your Comms

The single most common mistake new operators make is treating a 50-watt radio like a low-draw accessory. They see a fuse box, they see an "Accessory" slot, and they lazy-wire the radio into the nearest 12V source.

Do not do this.

To understand why, we have to look at the physics of your truck's electrical system.

The Theory: Dirty Power vs. Clean Power

Your vehicle's electrical system is a chaotic storm of electromagnetic noise.

- **The Alternator:** This is a generator that spins at thousands of RPM, creating electricity to charge your battery. It generates Alternating Current (AC) which is rectified into Direct Current (DC). This process is messy. It creates a "ripple" in the voltage - a high-pitched electronic noise that rises and falls with your engine RPM.

- **The Fuel Pump & Injectors:** High-speed electric motors and solenoids that click open and closed hundreds of times a second, sending spikes back down the power lines.

- **The LED Light Bars:** This is the silent killer of radio performance. Cheap off-road lights and USB chargers use "switching power supplies" that chop electricity up rapidly. They act like miniature radio jammers, spewing Radio Frequency Interference (RFI) everywhere.

If you wire your radio to the fuse box inside the cabin, you are tapping into a "dirty" circuit shared by the blower motor, the turn signals, and the fuel pump. You are asking your radio to listen for a weak signal (a whisper) while plugging its ears into a jackhammer.

The Symptom: "Alternator Whine." You press the gas pedal, and your radio emits a high-pitched *wheeeeeeee* sound that gets higher as you rev the engine. It makes you sound like an amateur, drives everyone on the channel insane, and masks incoming weak signals.

The Golden Rule: Direct to Battery

The cleanest power source in your vehicle is the **Battery Lead Acid Plate**. The battery acts as a massive chemical capacitor. It absorbs the ripples from the alternator and smooths out the spikes from the injectors. It is a buffer - a calm pool of water next to a raging river.

The Protocol:

1. **Wire Gauge Matters:** Do not use thin speaker wire. For a 50-watt radio, use **10 AWG or 12 AWG** automotive-grade wire. Thin wire causes "Voltage Drop." When you transmit, the radio demands 10-12 amps instantly. If the wire is too thin, the voltage drops below 11V, and the radio reboots or shuts down mid-sentence.

2. **Positive (+):** Run the red power wire from the radio directly to the Positive terminal of the vehicle's battery.

3. **Negative (-):** Run the black ground wire from the radio directly to the Negative terminal of the battery. If you must use a chassis ground point, **you must scrape the paint down to bare metal**. A bolt through paint is not a ground; it is a capacitor.

Note regarding modern vehicles with Battery Management Systems (BMS): Some newer trucks (F-150s, Toyotas, Broncos) have a sensor on the negative terminal clamp. This sensor tells the computer how much energy is leaving the battery. If you wire directly to the clamp, bypassing the sensor, the computer might get confused and fail to charge the battery correctly. In these cases, follow the negative cable from the battery to where it bolts to the frame (usually 6 inches away). Ground your radio *there*.

The Fuse: Your Fire Insurance

Every wire that leaves the positive terminal of your battery is a potential welding rod. If that wire chafes against the firewall and shorts out, it will glow red hot, melt the insulation, and set your carpet on fire.

Rule: The fuse protects the *wire*, not the radio. You must place a waterproof fuse holder on the positive wire **within 12 inches of the battery**, rated for your specific radio (usually 15A or 20A). If the wire shorts out anywhere between the battery and the radio (like where it passes through the firewall), the fuse pops, and your truck doesn't burn down. Do not rely on the fuse built into the back of the radio chassis or the one inline near the radio body. Those protect the radio from internal faults; they do nothing for a short in the engine bay.

Ferrite Chokes: The "Magic Beads"

Sometimes, even with direct wiring, you still pick up static or interference - often from your own accessories. This is where we use **Ferrite Chokes** (Mix 31 or Mix 43 snap-on cores).

The Physics: A ferrite choke is a chunk of magnetic ceramic. When you clamp it around a wire, it acts as a resistor for high-frequency energy (RF noise) but passes low-frequency energy (DC power) without issue.

The "Wattage is Water" Analogy: Imagine your power wire is a hose carrying water (electricity) to your radio. RFI noise is like air bubbles trapped in the water, causing sputtering. A ferrite choke is a filter that traps the bubbles but lets the water flow freely.

Deployment:

1. **Snap-on:** Buy 5mm or 7mm snap-on chokes.

2. **Loop it:** Don't just clip it on. **Loop the wire through the choke twice** if possible. The physics of inductance means that two turns through the core quadruples the noise-suppressing effectiveness.

3. **Location:** Place one choke on the power cable as close to the radio body as possible. Place another on the power cable as close to the battery as possible.

Result: A "black background" on your audio. Silence when no one is talking, allowing you to hear distant stations that would otherwise be buried in the noise floor.

9.2 Routing the Coax: The Arteries of the System

In Chapter 6, we mounted the antenna. Now we have to get the coaxial cable (the thick black wire) from the roof/bumper into the cab without destroying it.

Coax cable (usually RG-58, RG-8X, or LMR-240) is not a rope. It is a precision-engineered pipe consisting of a center copper conductor, plastic insulation, a braided metal shield, and an outer jacket. The spacing between the center wire and the shield must remain perfect.

The Crush Rule: If you pinch the cable hard enough to deform it (turn it from round to oval), you have ruined it. You have changed the "impedance" at that spot from 50 ohms to something else. This acts like a kink in a garden hose - the signal hits the kink and bounces back (SWR), heating up your radio instead of transmitting.

The "Pinch Point" Danger Zones

1. The Door Jam Guillotine

The lazy installer runs the coax through the door opening and slams the door shut on it. The rubber weather stripping might protect it for a week, but eventually, the metal door frame will work-harden the copper shield and snap it.

- **Verdict:** Never run coax through a primary door jam. It *will* fail, usually when you need it most.

2. The Tailgate Hinge

If mounting on a tailgate, you must route the cable alongside the factory wiring harness. Watch the "scissor action" of the hinge.

- **The Fix:** Use zip ties to secure the coax to the existing wire loom. Cycle the tailgate open and closed slowly multiple times. Watch the cable. Does it pull tight? Does it kink? It needs a "service loop" - a bit of slack that allows movement without tension.

3. The Hood Hinge

For "ditch light" mounts or lip mounts, you often route cable into the engine bay.

- **The Trap:** The hood hinge mechanism is a metal crushing machine.

- **The Fix:** Route the cable *behind* the hinge mechanism, securing it to the firewall cowl, ensuring it never touches the moving arms. Use split-loom tubing to protect the coax from engine heat.

Penetrating the Firewall: The Right Way

You need to get the power wire and the antenna coax from the engine bay into the cabin. The barrier is the Firewall - the steel plate separating you from the engine.

Method A: The "Daylight" Trick (Best) Look under the dashboard with a flashlight (up near the brake pedal). Look for the factory rubber boot where the main wiring harness passes through. Have a helper shine a bright flashlight from the engine side at that same rubber boot.

- **The Goal:** If you see light, or a nipple on the rubber boot, that is your path.

- **The Tool:** Use a wire coat hanger (straightened out) or a "fish tape." Tape your wire to the end of the coat hanger (use electrical tape, make a smooth, tapered cone so it doesn't snag).

- **The Move:** Poke the coat hanger carefully through the side of the rubber boot (do not cut the main wires!). Push it through until your helper grabs it in the engine bay. Pull the wire through.

- **The Seal:** If the hole is loose, dab a little RTV silicone sealant on it. Engine fumes (carbon monoxide) can enter the cabin through unsealed firewall holes.

Method B: The Drill (Last Resort) If you must drill a new hole:

1. **Check the other side:** Ensure you aren't drilling into the brake booster or a wiring loom.

2. **Drill Pilot:** Start small.

3. **Step Bit:** Use a "Unibit" step drill to make a clean hole.

4. **The Grommet: MANDATORY.** You cannot run a wire through a bare metal hole. The vibration will act like a saw, cutting through the insulation in 100 miles and shorting out the car. You *must* install a rubber grommet (available at any auto parts store) in the hole before passing the wire.

The Drip Loop

Water travels down wires. If your coax runs down from the roof and straight into a firewall grommet, rain will follow the wire, go through the grommet, and drip onto your feet.

- **The Fix:** Create a "Drip Loop." Before the wire enters the hole, bend it down into a 'U' shape and then back up into the hole. Water will run down to the bottom of the 'U' and drip off onto the ground, rather than climbing up into the cabin.

9.3 SWR (Standing Wave Ratio): The Engine Tune-Up

You have power. You have an antenna. You have coax. You are ready to transmit, right?

STOP.

If you key the mic now, you might destroy your new $200 radio. You must check the **SWR**.

The Physics of the "Parking Brake"

Radio Frequency (RF) energy flows out of your radio, down the coax, and into the antenna. The antenna's job is to radiate that energy into the air.

- **In a perfect world (SWR 1.0:1):** 100% of the energy leaves the antenna.

- **In the real world:** The antenna is never perfectly tuned to the exact frequency you are using. Because of this mismatch, some of the energy hits the end of the antenna and *bounces back* down the coax, returning to the radio.

The Analogy: Imagine you are spraying a fire hose (the radio) through a nozzle (the antenna).

- If the nozzle is open, water flies out.

- If you put your thumb over the nozzle (bad tuning), the water sprays back in your face.

- If you plug the nozzle completely, the water pressure backs up and bursts the hose or blows up the pump.

High SWR (Standing Wave Ratio) is that back-pressure.

- **Heat:** The energy that comes back has to go somewhere. It turns into heat inside your radio's "final transistors."

- **Damage:** If the SWR is too high (above 3.0), the radio will overheat and burn out its transmitter components in seconds.

- **Performance:** Even if it doesn't break, High SWR means your signal isn't getting out. An SWR of 2.5 means you might only be transmitting 30 watts instead of 50. You are driving with the parking brake on.

The Tool: The SWR Meter

You need an SWR meter covering the UHF (GMRS) frequencies (462-467 MHz). A standard CB radio meter *will not work* (wrong frequency range).

- **Budget:** Surecom SW-102 (Digital, easy). Highly recommended for the glovebox.

- **Pro:** NanoVNA (Complex, but powerful). Shows you the antenna curve across the whole band.

- **Analog:** Daiwa CN-901 (Old school, reliable). No batteries required, but bulky.

The Tuning Procedure (The "Measure Twice, Cut Once" Protocol)

Setup:

1. Drive your vehicle to an open area. **Do not** tune inside a garage. Metal walls, concrete ceilings, and nearby cars will reflect the signal and give you false readings. You need at least 20 feet of clear space.

2. Close all doors and windows. Open doors change the ground plane shape.

3. Connect the Meter:

  ○ Radio -> Patch Cable -> Meter Input (TX).

  ○ Meter Output (ANT) -> Coax -> Antenna.

The Test:

1. Set your radio to a middle channel (GMRS Channel 19).

2. Set radio power to **Low** (to protect it if the SWR is terrible).

3. Key the microphone (Press PTT).

4. Read the meter.

Interpreting the Number:

- **1.0 to 1.5: Excellent.** Do not touch anything. You are a wizard.

- **1.5 to 2.0: Good.** Acceptable for daily use.

- **2.0 to 2.5: Marginal.** You are losing power, and the radio will run warm. Try to improve it.

- **2.5 to 3.0+: DANGER.** Do not transmit. Something is broken or badly mistuned. Check for a short in the coax or a bad ground.

How to Tune (Adjusting the Length)

An antenna is a bell. We want it to ring at a specific pitch (462 MHz). We adjust the pitch by changing the length of the metal whip.

- **If SWR is lower on Channel 1 than Channel 22:** The antenna is **too long**. (It resonates at a lower frequency).

  ○ *Action:* Loosen the set screw (grub screw) on the antenna whip. Pull the metal whip out. **Mark the current depth with a Sharpie** so you know where you started. Snip off **1/8th of an inch** (3mm) using linesman pliers or a dremel. Re-insert. Tighten. **Step away from the vehicle.** Your body reflects RF signals. If you lean over the antenna to watch the meter, you will get a false reading. Test again.

- **If SWR is lower on Channel 22 than Channel 1:** The antenna is **too short**. (It resonates at a higher frequency).

   ◦ *Action:* Raise the whip up in the mount if possible. If you already cut it too short... congratulations, you just bought a new replacement whip ($15). This is why we cut in tiny increments. You can't put metal back on.

The "Ground Plane" Factor: If you cannot get the SWR below 2.5 no matter what you do, you likely have a **Ground Plane** issue. The antenna whip is only half the antenna. The other half is the "mirror image" reflected by the metal surface under it.

- **Diagnosis:** Your antenna doesn't have enough metal under it (e.g., you mounted it on a plastic roof rack, a composite camper shell, or a tube bumper).

- **The Fix:** You must run a ground strap (braided wire) from the antenna mount to the vehicle's chassis, or add a metal plate/disc (at least 6-8 inches diameter) under the antenna base.

Action Step: The "Install Inspection" (The 10-Point Check)

You've wired, routed, and tuned. Before you declare the rig "Mission Ready," execute this inspection. Loose wires start fires, and vibration finds every weakness.

The Checklist:

1. [ ] **Battery Connections:** Are the ring terminals tight on the battery? Can you rotate them by hand? (They should be rock solid).

2. [ ] **Fuse Placement:** Is the fuse within 12 inches of the battery? Is the cap on the fuse holder sealed against water/mud?

3. [ ] **Wire Chafe (Engine Bay):** Follow the red wire. Does it touch the hot engine block, the exhaust manifold, or sharp metal edges? Zip tie it away from heat and abrasion. Use split-loom tubing for extra armor.

4. [ ] **Grommet Integrity:** Check where the wires enter the firewall. Is the grommet seated? Is the silicone dry?

5. [ ] **Interior Routing:** Look under the dash. Are wires dangling near the brake or gas pedals? (Zip tie them up high! A wire catching your shoe can cause an accident).

6. [ ] **Coax Kinks:** Check the coax run. Are there any sharp 90-degree bends or pinched spots? Ensure the "Minimum Bend Radius" is respected (usually the size of a soda can).

7. [ ] **Connector Tightness:** Check the PL-259 connector on the back of the radio. Hand-tighten, then give it a tiny nip (1/8 turn) with pliers. Vibrations loosen these over time.

8. [ ] **Radio Mount:** Grab the radio body. Shake it violently. If it moves, the mounting screws are loose. Off-road driving creates earthquake-level vibration.

9. [ ] **SWR Verification:** Final test. Is SWR under 2.0 on Channel 1 and 22?

10. [ ] **Audio Check:** Call for a radio check. Ask specifically: "How is my signal? Is there any whine or static when I rev the engine?"

The "Shake-Down" Run: The final test is the road. Go find a washboard dirt road. Drive it for 20 minutes. Stop. Check the mount. Check the SWR again. If your antenna mount is loose, washboards will find it. If your power connection is intermittent, washboards will find it. If nothing fell off and nothing is smoking, you are cleared for operations.

Chapter 10

Troubleshooting

The Visual Command Flowcharts

OBJECTIVE: DIAGNOSE AND FIX communication failures in under 60 seconds using binary logic.

Key Deliverable: Three field-ready diagnostic trees for your glovebox.

Introduction: Logic, Not Magic

You are on the trail. It's late afternoon, the shadows are getting long, and the "check engine" light of the lead vehicle just flickered. You pick up the mic to coordinate a stop. You press the button. Nothing happens. You press it harder (because clearly, hydraulic pressure fixes electronics). Still nothing.

At this moment, the "Tactical Dad" usually splits into two distinct, unhelpful personalities:

1. **The Magician:** This operator believes that if he randomly mashes buttons, twists knobs back and forth, and shakes the radio like a Polaroid picture, he will accidentally stumble upon the solution. He is hoping for magic. He usually ends up changing a setting that was actually correct, compounding the problem.

2. **The Defeatist:** This operator tries once, fails, and immediately spirals. He throws the radio on the passenger seat, declares the entire GMRS system "junk," and drives the rest of the day in angry silence, stewing over the money he spent.

Both approaches are failures of command. They are emotional reactions to a mechanical problem.

Radio problems are rarely mysterious. They are mechanical, electrical, or procedural. They follow the laws of physics, not Murphy's Law. In the *Visual Command System*, we do not guess. We trace the signal path. We ask "Yes/No" questions until we find the failure point.

Troubleshooting is simply the art of eliminating variables.

- If the radio turns on, the battery is good.

- If it hisses when you drop the squelch, the speaker is good.

- If it receives but can't transmit, the antenna is likely physically intact, but the settings are wrong.

This chapter replaces the frantic "Magician" with the calm, methodical **Diagnostician**.

10.1 The "I Can't Hear You" Flowchart (Receiver Failure)

The Scenario: You can see the lead vehicle through your windshield. You can see the driver talking into his hand mic, gesturing wildly. Your radio is dead silent. You feel isolated, and that isolation breeds anxiety.

The Logic: The signal is leaving *their* truck (we assume) but being blocked before it hits *your* eardrum. We must check the "Gates" in order, from the easiest (human error) to the hardest (hardware failure).

Step 1: The "Idiot Check" (Volume & Power)

Do not skip this. Do not be offended. 30% of all radio failures on the trail are simply the volume knob getting bumped against a hip (HT) or a center console (Mobile).

- **The Tactile Check:** Turn the knob all the way down until it clicks off. Then turn it on and halfway up.

- **The Auditory Check:** Listen for the startup beep or voice prompt ("Channel Mode!"). If you don't hear the beep, the radio isn't making sound.

- **Visual Command:** Look at the screen. Is it lit?

  - *HT Fix:* If the screen is dead, remove the battery and wipe the contacts with your thumb. Re-seat it firmly.

○ *Mobile Fix:* Check the inline fuse near the battery. A loose fuse holder can disconnect power on bumpy roads.

Step 2: The Channel Match

GMRS channels are standardized by the FCC, but channel *names* and memory slots are not. This is a massive source of confusion when mixing radio brands (e.g., Midland vs. Wouxun vs. Baofeng).

- **The Problem:** Your buddy says, "Go to Channel 5." He means the 5th slot in *his* radio's memory, which he custom-programmed as the "Rubicon Repeater" (462.675 MHz). You go to the standardized GMRS Channel 5 (462.6625 MHz). You are now existing in parallel universes.

- **The Fix:** Confirm the **Frequency**, not just the Channel Number.

  ○ *Command:* "Lead, this is Chase. Confirm you are on frequency 4-6-2 point 6-6-2-5?"

  ○ If they don't know how to check their frequency, you must switch to a "Universal Standard" channel like Channel 16 (462.575) to re-establish comms.

Step 3: The "Privacy Tone" Mismatch (The Silent Killer)

This is the #1 cause of "I can see you talking but I can't hear you." It happens when someone accidentally enables a "Receive Tone" (RX CTCSS/DCS) on their radio.

- **The Physics:** Remember, a Privacy Tone is an electronic key. If your radio is set to expect "Tone 100.0 Hz," it creates a locked door. It will *only* open that door if the incoming signal carries the 100.0 Hz key. If the sender is not transmitting that tone (or is sending a different one), your radio hears the signal but refuses to open the speaker.

- **The Diagnostic:** Press the **MONITOR** (Moni) button on the side of your radio.

  ○ *What this does:* It forces the Squelch gate open and temporarily ignores all privacy tones. You will hear loud white noise (static).

  ○ *The Test:* Hold 'Monitor' while the other person talks.

  ○ *Result A:* **You hear them clearly through the static.**

- *Verdict:* Your settings are wrong. You have a Receive Tone set that shouldn't be there. Go to Menu -> Rx Tone -> Set to OFF (or 0).

  ○ *Result B:* **You still hear only static while they talk.**

    - *Verdict:* The issue is not the Tone. It is likely Frequency (you aren't on the same channel) or Range (they are too far away).

Step 4: The Squelch Trap

Squelch is the gatekeeper that keeps the background cosmic radiation and electrical noise quiet. If the gate is set too high (too strong), it won't open for a weak signal.

- **The Scenario:** You are separated by a large hill. The signal is weak. Your squelch is set to Level 5 (Factory Default). The weak signal hits the gate but isn't strong enough to push it open.

- **The Check:** Look at your screen. Is the **"RX" (Receive) icon** lighting up green (or showing signal bars), but no sound is coming out? This indicates the radio *hears* something but the Squelch is suppressing it.

- **The Fix:** Lower your Squelch setting (SQL). Drop it to 1. This makes the "gate" lighter, allowing weaker signals to pass through.

Step 5: The "Headset" Ghost

Did you have an earpiece or hand-mic plugged in earlier? Did you drop your HT in a puddle?

- **The Problem:** The headphone jack is a mechanical switch. When you insert a plug, it physically disconnects the main speaker. Sometimes, dirt, lint, or water gets jammed in the jack, tricking the radio into thinking an accessory is still plugged in. It routes audio to a non-existent headset.

- **The Fix:** Plug and unplug the accessory jack 3-4 times rapidly to clear debris. Blow it out with compressed air (or a sharp breath).

Step 6: The "RFI Wall" (The Overland Special)

This is unique to modern modified vehicles. You might be jamming *yourself*.

- **The Scenario:** You hear constant static, or your range has dropped to 50 feet.

- **The Test: Turn off your engine.** Turn off your LED light bars. Unplug your cheap USB chargers, Dash Cams, and AC Inverters (which are notoriously noisy in the RF spectrum).

- **The Reveal:** If the radio suddenly comes to life and you can hear your group, your vehicle is generating Radio Frequency Interference (RFI). Cheap LED drivers and USB buck converters scream across the UHF spectrum, drowning out incoming signals.

- **The Fix:** Buy high-quality electronics (e.g., ferrite-shielded lights) or accept that you cannot use the radio and the lights simultaneously.

The "I Can't Hear You" Decision Tree

```
START
  |
  +-> [1] Is Radio ON and Volume UP? (Did you hear the beep?)
  |       NO  -> Check Battery / Fuse.
  |       YES -> Continue.
  |
  +-> [2] Do Frequencies Match? (Ignore Channel Names)
  |       NO  -> Match Frequencies (e.g., 462.5625).
  |       YES -> Continue.
  |
  +-> [3] Press & Hold MONITOR Button. Do you hear them now?
  |       YES -> TONE MISMATCH. Turn off your Rx Tone (CTCSS/DCS).
  |       NO  -> Continue.
  |
  +-> [4] Is the Green Rx Icon lighting up?
  |       YES -> Check Squelch (Lower it to 1) or check for "Headset Ghost."
  |       NO  -> Continue.
  |
  +-> [5] Turn off Engine/Lights. Did reception improve?
  |       YES -> RFI Interference. Your truck is the problem.
  |       NO  -> They are likely out of range or not transmitting.
```

10.2 The "You Can't Hear Me" Flowchart (Transmitter Failure)

The Scenario: You are talking, giving directions, telling jokes. But nobody is answering. You feel like a ghost. You suspect your radio isn't "getting out," or perhaps they are just ignoring you.

The Logic: The signal is dying somewhere between your voice box and the antenna tip.

Step 1: The PTT (Push-To-Talk) Discipline

Are you pushing the button *before* you speak, or *as* you speak?

- **The Problem:** Digital radios, and especially Repeaters, take a few milliseconds to "wake up" and open the audio path. If you say "Rock in the road," but you push the button on the "o" in Rock, they hear "...ck in the road." This leads to confusion and constant "Say again?" requests.

- **The Fix:** The "One-Mississippi" Rule. Push. Pause (inhale). Talk. This ensures the carrier wave is established before your voice modulation begins.

Step 2: The Microphone Connection

On mobile units, the hand mic connection is the physical weak link. It gets pulled, twisted, stepped on, and yanked.

- **The Diagnostic:** While holding the PTT button, wiggle the connector where it enters the radio body. Watch the **Red TX Light**.

  - Does it flicker?

  - Does it go out?

- **The Fix:** Unplug it. Check for bent pins. Blow it out. Re-seat it firmly until it clicks. On HTs (Handhelds) with the 2-pin Kenwood connector, you often have to press *very* hard to get the plug fully seated past the weather seal.

- **Check Mic Gain:** Ensure your "Mic Gain" setting in the menu hasn't been inadvertently lowered. If set too low, your modulation will be so quiet that others will think you aren't transmitting.

Step 3: TX Power (High vs. Low)

Are you whispering when you should be shouting?

- **The Problem:** You are trying to hit a repeater 20 miles away, or a vehicle on the other side of a ridge. Your radio is set to "L" (Low Power / 5 Watts) instead of "H" (High Power / 50 Watts).

- **The Fix:** Look at the screen for a small "L" or "H". Go to Menu -> TX Power -> Switch to High.

  - *Regulatory Note:* Remember that Channels 8-14 are legally restricted to Low Power (0.5W) on all radios. If you are on Channel 12, you *cannot* switch to High. If you need range, move the convoy to Channel 16.

Step 4: Repeater Offsets (The "Input" Error)

This is the classic Rookie Repeater Mistake.

- **The Scenario:** You are trying to talk on a Repeater (e.g., The "700"). You can *hear* the repeater traffic perfectly. But when you talk, nobody hears you.

- **The Physics:** Repeaters are "Duplex" machines. They listen on one frequency (Input) and talk on another (Output). Your radio must be programmed to *shift* its transmission to the Input frequency automatically when you press PTT. This is called the "Offset" (+5.000 MHz).

- **The Diagnostic:** Look closely at your screen's frequency display. Press the PTT.

  - *Rx Frequency (Listening):* 462.700

  - *Tx Frequency (Pressing PTT):* **Does it change to 467.700?**

  - If the number *stays* 462.700 when you push the button, you are talking on the Output. You are shouting at the repeater's mouth instead of its ear.

- **The Fix:** Enable the "+5M", "OFFSET", or "RPT" function in the menu (depending on your radio brand).

Step 5: Repeater Input Tones

The Repeater "Password."

- **The Scenario:** You have the offset correct (the frequency shifts). You key up. You hear nothing back (no "ker-chunk" or "tail").

- **The Problem:** Most repeaters are kept private (or semi-private) to prevent interference. They require a specific PL Tone (e.g., 141.3 Hz) to "wake up." If you aren't sending that tone, the repeater ignores your signal entirely.

- **The Fix:** Verify the **Tx Tone** (Transmit CTCSS) in your menu matches the repeater's requirement. *Note: Many regional repeater networks use a standard "Travel Tone" of 141.3 Hz. If you are guessing, try that first.*

The "You Can't Hear Me" Decision Tree

```
START
  |
+-> [1] Does the Red 'TX' Light turn on when PTT is pressed?
  |      NO  -> Check Mic Connection (Push hard) / Check Battery.
  |      YES -> Continue.
  |
+-> [2] Are you on the correct Channel/Frequency?
  |      NO  -> Match Group.
  |      YES -> Continue.
  |
+-> [3] Are you trying to use a Repeater?
  |      NO (Simplex) -> Check Range. Switch to High Power. Try moving vehicle 5 feet.
  |      YES (Repeater) -> Continue below.
  |
      +-> [3a] Does frequency shift +5MHz when PTT pressed?
      |      NO  -> Wrong Setting. Enable Offset (+5.000).
      |      YES -> Continue.
      |
      +-> [3b] Is the correct Tx Tone programmed?
             NO  -> Enter Tone (Try 141.3 if unknown).
             YES -> Repeater might be down or out of range.
```

10.3 The "Radio is Hot/Smells" Flowchart (Hardware Failure)

The Scenario: You pick up the hand mic and the radio body burns your hand. Or, you smell the distinct, sharp scent of "Magic Smoke" (burning ozone/plastic). Or, your transmission starts strong but cuts out or gets "robotic" after 10 seconds.

The Logic: Energy is not leaving the antenna. It is bottling up inside the radio and turning into heat. This is an immediate hardware emergency.

Step 1: The "Touch Test"

Mobile radios get warm. That is normal. They act as their own heatsinks, dissipating 50 watts of power.

- **Normal:** Warm coffee mug. Uncomfortable but holdable.

- **Danger:** Hot stove. You reflexively pull your hand away.

- **Action:** If it is "Hot Stove" hot, **STOP TRANSMITTING IMMEDIATELY.** Turn the radio off. Let it cool for at least 20 minutes.

Step 2: The SWR Check (Reflected Power)

Refer back to Chapter 9. Heat is almost always caused by High SWR (Standing Wave Ratio).

- **The Physics:** The antenna is a nozzle. If the nozzle is clogged (bad tuning, broken wire), the energy hits the clog and bounces back down the cable. It returns to the radio and cooks the final transistors.

- **The Field Fix:**

    a. **Visual Inspection:** Is the antenna still there? (Did a low tree branch snap it off?) Is the coil bent?

    b. **Cable Trace:** Is the coax cable pinched in the door jamb? Is the connector loose on the back of the radio?

    c. **Meter Check:** If you have an SWR meter, check it. If SWR is > 3.0, **DO NOT TRANSMIT.** You are dead in the water until you fix the antenna system.

Step 3: The "Duty Cycle" Abuse

GMRS radios are not chat rooms. They are not designed for "podcasting."

- **The Rating:** Most mobile radios have a "Duty Cycle" of 10-20%. This means for every 1 minute of talking, you need 4 minutes of listening/cooling.

- **The Abuse:** If you (or your kids) hold the button down for 5 minutes straight telling a story or reading a book, the radio will overheat and engage thermal shutdown.

- **The Fix:** Brief brevity. Keep transmissions under 30 seconds. If you have a "talker" in the group, gently remind them: "Break for cooling."

Step 4: The Voltage Drop (The "Brownout")

Is the radio cutting out, rebooting, or dimming when you press PTT?

- **The Cause:** Bad wiring. When you transmit, the radio tries to pull 10-12 Amps of current. If you used thin wire, or if your ground connection is loose/corroded, the pipe is too small. The voltage drops from 13.8V to 9V. The radio detects the drop and shuts off to protect itself.

- **The Fix:** Check your fuse holder (is the plastic melted?). Check your battery terminals. Ensure you wired directly to the battery, not a cigarette lighter plug.

The "Hardware Crisis" Decision Tree

```
START
 |
+-> [1] Is there smoke or a burning smell?
 |      YES -> KILL POWER IMMEDIATELY. Pull the fuse. Do not turn on.
 |      NO  -> Continue.
 |
+-> [2] Is the radio too hot to touch?
 |      YES -> Stop Transmitting. Let cool for 20 mins. Check Duty Cycle.
 |      NO  -> Continue.
 |
+-> [3] Does the radio reboot/dim when you press PTT?
 |      YES -> Check Battery Voltage / Ground Wire / Fuse Holder.
 |      NO  -> Continue.
 |
+-> [4] Check SWR / Antenna
        Is the antenna broken? Is the coax pinched?
        If SWR is High -> Fix Antenna before transmitting.
```

Action Step: The "Broken Arrow" Drill

Troubleshooting charts are useless if you can't read them because you are panicking or frustrated. You need muscle memory.

The Exercise: Next time you are out with your group (or just in the driveway with the kids), initiate a "Broken Arrow" drill.

1. **Sabotage:** Have your spouse take your radio and change *one* setting without you looking.

   ○ Turn on a random Receive Tone.

   ○ Turn Squelch to Max.

   ○ Change the Offset.

   ○ Unscrew the antenna connector halfway so it's loose.

2. **The Test:** Get in the truck. Close the door. Try to call them.

3. **The Solve:** Do not guess. Use the Flowcharts mentally.

   ○ *"I have power."*

   ○ *"I am on the right channel."*

   ○ *"They are talking but I can't hear them."*

   ○ *"Press Monitor... ah, I hear them now. It must be a Tone mismatch."*

   ○ *"Checking Tone settings... yes, Tone is on. Turning it off."*

Time Standard: You should be able to diagnose and fix a "Soft Fail" (settings issue) in under 60 seconds.

By mastering these trees, you stop being the guy who bangs on the dashboard when things don't work. You become the guy who fixes the comms, calms the group, and gets the convoy moving again. You become the **Visual Command**.

Chapter 11

Advanced Operations

Scanning & Monitoring

OBJECTIVE: TRANSFORM YOUR RADIO from a passive telephone into an active intelligence-gathering tool.

Key Deliverable: A "Sentinel" configuration that watches the weather, the convoy, and the threat environment simultaneously.

Introduction: The Intelligence Officer

Up to this point in the book, we have treated the radio primarily as a **Tactical Transmitter** - a device for you to send commands to your family or convoy. You push the button, you speak, they listen. It is a tool of output.

But in the *Visual Command System*, transmission is only half the battle. The other half - and often the more critical half - is **Reception**, or more accurately, **Intelligence Gathering**.

Imagine you are driving into a box canyon in southern Utah. The sky is turning a bruised purple, signaling a pressure drop. Your convoy is chatting happily on Channel 16 about lunch. But what is happening five miles ahead? Is there a flash flood warning issuing for this specific slot canyon? Is there a logging crew blocking the exit route on Channel 20? Is there a rescue operation happening on the local repeater, implying the trail is closed?

If you sit silently on Channel 16, waiting for someone to call you, you are driving blind. You are reactive, not proactive.

You need to become the **Intelligence Officer**. You need to configure your radio to act as an electronic sentry, sweeping the airwaves for threats, weather changes, and logistical information while you keep your eyes on the trail. This is where we move beyond "Push-to-Talk" and master the art of Scanning, Dual Watch, and Information Dominance.

11.1 Scanning Strategies: The Electronic Sentry

Scanning is simply the radio's ability to cycle through a list of channels rapidly, stopping only when it hears a voice. It sounds simple, but most users get it wrong. They press "Scan" and get blasted by white noise, or they get stuck listening to a drive-thru order at a Taco Bell three miles away, missing their own team's check-in.

To scan effectively, you must master three variables: the **Scan List**, the **Scan Mode**, and the **Tone Scan**.

The Scan List: Curating Your Feed

Your radio likely has 200+ memory channel slots. You do not want to scan all of them. If you scan every possible frequency (GMRS, FRS, MURS, Marine, Local Police), the cycle time - the time it takes to get back to Channel 1 - might be 30 seconds. In those 30 seconds, your lead driver could call out "Rockfall!" and you would miss it because your radio was listening to a garbage truck on Channel 2.

Rule: A Scan List should be mission-specific. You don't take a snow shovel to the desert; don't put snow plow frequencies in your desert scan list.

- **The Convoy List (Group A):** This is tight and fast. It includes only the Convoy Channel (e.g., Ch 16), the Alternate (Ch 20), and the primary Emergency Repeater. The cycle time is under 1 second. You will never miss a word.

- **The Area List (Group B):** Includes all GMRS channels, plus local Sheriff/Fire frequencies (listen-only) relevant to the specific county you are traversing. This is for situational awareness when the convoy is parked or quiet.

How to "Skip" (The Noise Filter): When programming your radio (especially in CHIRP software), every channel has a "Skip" column. This is your primary filter.

- **"S" or "Skip":** The scanner ignores this channel completely.

- **"Off" or Blank:** The scanner includes this channel in the cycle.

Strategy: Permanently "Skip" the FRS/GMRS hybrid channels (8-14). These channels are legally restricted to low power (0.5 watts) and have fixed antennas. They are the playground of blister-pack radios sold at big-box stores. They are usually clogged with children playing tag or families coordinating parking at the mall. They are noise, not signal. Remove them from your scan list immediately to increase your scan speed.

"Nuisance Delete": The Field Fix

Here is the scenario: You are scanning your Area List. You are looking for emergency traffic or trail reports. Suddenly, the scan stops on Channel 22. It's a construction crew fixing a bridge five miles away. *"Bring the loader... back it up... beep beep beep..."* The radio stays locked on them. You press "Scan" to resume. It cycles for 2 seconds, hits Channel 22 again, and stops. *"Yeah, dump that load here."*

You are now in **Scan Hell**. You can't hear your team because the construction crew is dominating the airwaves, and you can't walk away from the radio.

Do not go into the menu to delete the channel. That is too slow and dangerous while driving. **Use "Nuisance Delete."**

On almost all modern GMRS radios (Wouxun, Midland, Baofeng, BTech), there is a shortcut feature specifically for this. When the scanner stops on a channel you don't want to hear:

- **The Move:** Press and hold the "*" **(Star)** key or the designated **Side Key** (check your manual).

- **The Result:** The radio beeps and *temporarily* removes that channel from the active scan list. It will not stop there again until you cycle power.

- **The Reset:** When you turn the radio off and on again, the channel is restored to the list.

Visual Command Logic: This allows you to aggressively filter your feed in real-time. If it's not relevant to the mission, Nuisance Delete it. Clear the deck so you can hear what matters.

Scanning Modes: CO vs. TO vs. SE

This setting confuses everyone because the acronyms are non-intuitive. It dictates *how* the radio behaves when it finds a signal. Choosing the wrong one breaks your flow.

1. **TO (Time Operation):** The radio stops on a signal for a set time (e.g., 5 seconds), then resumes scanning *even if the person is still talking*.

   ○ *Scenario:* You hear: "Be advised, there is a massive washout at mile marker..."

   ○ ***Scan Resumes***

   ○ You missed the location.

   ○ *Verdict:* **Useless.** Avoid this setting.

2. **SE (Search/Terminate):** The radio stops on a signal and stays there forever. It stops scanning completely.

   ○ *Scenario:* You find a signal. You listen. The conversation ends. You drive for 20 minutes, realizing too late that your radio stopped scanning and you are now listening to silence on a dead channel.

   ○ *Verdict:* **Bad.** This defeats the purpose of a sentry.

3. **CO (Carrier Operation):** The radio stops when it hears a voice. It stays there as long as the voice is talking. Once the voice stops (and a 2-second buffer passes), it resumes scanning.

   ○ *Scenario:* You hear the full washout report. The transmission ends. The radio waits 2 seconds for a reply. Hearing none, it goes back to patrolling the frequencies.

   ○ *Verdict:* **The Standard.** Set your radio to CO (or "Busy" on some models). This mimics natural listening.

Priority Scanning: The "Check-In"

Some advanced radios (Wouxun KG-1000G, Midland MXT500) offer **Priority Scanning**. This is a game-changer for convoy leaders.

• **The Setup:** You set Channel 16 as your "Priority Channel."

- **The Action:** You start scanning Channels 1-22.

- **The Logic:** Even if the scanner stops on Channel 5 to listen to hikers, the radio will microscopically "check" Channel 16 every 0.5 seconds. If someone talks on Channel 16, the radio brutally cuts off the hikers and switches to Channel 16.

- **The Benefit:** You can monitor the world without fear of missing your primary team.

Tone Scanning: The Key to the City

You are in a new town. You scan and find a repeater on Channel 20 (462.675). You hear locals chatting. You try to talk to them, but they can't hear you. Why? You are missing the 'Privacy Tone' (CTCSS/DCS).

Instead of guessing random numbers, use **Tone Scan**.

1. **Tune** to the busy channel (Ch 20).

2. **Menu:** Find 'SC-CTC' or 'Tone Scan'.

3. **Activate:** The radio will listen to the incoming voice and cycle through all possible tones.

4. **Lock:** When it matches the tone (e.g., 141.3), the screen will flash 'CTC-SS 141.3' and stop.

5. **Save:** Save that tone to your Transmit setting. You now have the key to the city.

NOAA Weather Alerts: The Voice of God

Every "Overland Adventurer" needs to monitor the weather. Mountains generate their own weather systems. A sunny morning can turn into a hail-spiking thunderstorm in 20 minutes, turning a dry wash into a deadly river.

Your GMRS radio likely has 7 dedicated **NOAA Weather Channels** (frequencies 162.400 through 162.550 MHz). These are broadcast by the National Weather Service (NWS). They are automated, robotic voices reading the forecast on a loop.

The "Monitor" vs. "Alert" Dilemma:

- **Monitor Mode:** You tune to the weather channel manually and listen to the robot. Useful for morning briefings while making coffee.

- **Alert Mode:** You put the radio in "Weather Scan" or "Alert Standby." The radio is silent. It consumes very little power. However, if the NWS issues a Severe Thunderstorm Warning or Tornado Warning, they transmit a specific **1050 Hz Tone**. Your radio detects this tone and sounds a siren alarm (like a smoke detector), then opens the audio so you can hear the warning.

The Protocol: When you get to camp, do not turn your radio off.

1. Switch to the local NOAA channel that comes in clearest.

2. Activate "WX Alert" mode.

3. Leave the radio on overnight (connected to the Aux battery or a solar generator).

If a flash flood is coming down the canyon at 3:00 AM, that radio is your only warning system. It is the sentry that never sleeps.

11.2 Dual Watch: The Two-Eared Operator

Scanning is great for finding random traffic. But what if you have two specific, high-priority channels?

1. You need to listen to **Channel 16** (The Convoy) so you don't miss a turn.

2. You need to listen to **Channel 20** (The Scout) who is two miles ahead checking the bridge integrity.

If you are on Ch 16, you miss the Scout. If you are on Ch 20, you miss the Convoy. If you Scan, you might miss the start of a sentence.

Enter **Dual Watch** (often labeled TDR, DW, or Dual Wait).

The Concept: Time-Division Multiplexing

Most GMRS radios have one receiver. They cannot *truly* listen to two frequencies at the exact same nanosecond. However, Dual Watch switches the receiver back and forth between Frequency A and Frequency B roughly 5 to 10 times per second.

- Check A... Check B... Check A... Check B...

To the human ear, it sounds instant. If someone talks on A, the radio locks on A. If someone talks on B, it locks on B.

The Setup: A/B Configuration

On your radio screen, you usually have a Top Line (A) and a Bottom Line (B).

1. **Set Line A:** Channel 16 (The Convoy Main).

2. **Set Line B:** Saddleback Repeater (Emergency Comms / Scout).

3. **Activate:** Turn on "TDR" or "Dual Watch" in the menu. An icon (usually "DW" or "S") appears.

The Synchronization Trap

There is a flaw in Dual Watch: **Simultaneous Transmission.** If the Scout talks on Channel 20 *at the exact same moment* the Convoy talks on Channel 16, the radio will lock onto whichever signal is split-second faster or stronger. You will miss the other one completely.

- *Mitigation:* This is why "Radio Discipline" and brevity are important. If everyone chatters endlessly, Dual Watch fails. Keep messages short to leave airtime for the other channel to break through.

The Danger Zone: PTT Priority Here is where people get in trouble socially and tactically. You are listening to both A and B. You hear the Scout talk on Line B. You grab the mic and press the button to reply.

- *Which channel does your radio transmit on?*

This depends on a menu setting called **"TX Priority"** (Transmit Priority).

- **Option 1: Selected (A/B):** The radio transmits on whichever line has the "Arrow" or is highlighted on the screen.

  ○ *Risk:* You might be listening to B, but the arrow is parked on A. You press the button and blast your reply to the Convoy (A) instead of the Scout (B). You say "Bridge looks sketchy," and the Convoy on Ch 16 gets confused because they aren't at the bridge yet.

- **Option 2: Busy (Last Active):** The radio automatically transmits on the channel that *last received a signal*.

  - *Benefit:* If the Scout talks on B, and you press PTT within 5 seconds, you reply on B automatically.

  - *Risk:* If the Convoy (A) chips in right after the Scout stops talking, your reply might switch to A mid-thought.

The Visual Command Recommendation: Set TX Priority to **"Selected" (Manual)**. Train yourself to look at the screen before you press the button. It creates a deliberate OODA loop:

1. "I hear traffic on B."

2. "Look at screen."

3. "Press 'A/B' button to move arrow to B."

4. "Press PTT."

It adds one second to your reaction time, but it prevents "Cross-Channel Pollution," where you accidentally scream at the wrong group of people.

11.3 Digital/Data (The Future)

Historically, GMRS was strictly analog. Just voice waves modulating a carrier signal. But in 2017 and 2021, the FCC updated the rules. We are now allowed to send short data bursts (limited to seconds) on GMRS channels.

This has opened the door for **Text Messaging** and **GPS Location Sharing**.

The Hybrid Radios (BTech GMRS Pro / Motorola T800)

Radios like the **BTech GMRS Pro** or the **Motorola T800** are "App-Centric" radios. They connect to your smartphone via Bluetooth.

How it Works:

1. **The Input:** You open the companion app on your phone. You see a map. You type a text message: *"Camp site found. 34.556, -112.443."*

2. **The Transfer:** The phone sends the data to the radio via Bluetooth.

3. **The Burst:** The radio converts that data into a screeching audio burst (like an old dial-up modem or a fax machine) and blasts it over the GMRS frequency.

4. **The Decode:** The receiving radio hears the screech, decodes it, and pops the text message up on *their* phone screen or radio display.

The Analog Annoyance (Etiquette Warning): To you, it's a silent text. To everyone else on Channel 16 using a normal radio, it sounds like a screeching modem from 1999. **Rule:** Warn the group before sending data. 'Sending GPS coords, stand by for data burst.' Do not spam data on a busy voice channel.

The Tactical Advantage: Silent Command

Why would you text when you can talk? Why complicate a simple tool?

1. **Coordinate Accuracy:** Reading GPS coordinates over voice is a nightmare. *"Three four... did you say four or five?... point five five..."* It is prone to error. Sending a data burst is 100% accurate. The recipient just clicks the map pin and their navigation starts.

2. **Silence:** You are hunting. You are observing wildlife. Or you are in a security situation where you do not want to make noise. Texting allows for silent coordination without breaking noise discipline.

3. **Persistence:** If you call someone on voice and they are in the bathroom or changing a tire, they miss the message. A text message waits on their screen until they return. It creates a log.

The Limitation: The Walled Garden

The catch is compatibility. Analog voice is universal; a Midland can talk to a Wouxun. **Data is proprietary.**

- A BTech GMRS Pro cannot send a text to a Motorola T800. They speak different digital languages (modulation schemes).

- A Garmin Rino cannot send location data to a Baofeng.

The Verdict: For a tight-knit family or a dedicated squad, Digital GMRS is a superpower. It provides a "Mini-Internet" completely off-grid. But for a general convoy with mixed gear, stick to Analog Voice (Channel 16). Use Data as your "Inner Circle" comms layer.

Battery Warning: Using Bluetooth and GPS features drains your radio battery significantly faster. If you are running digital, bring spare batteries.

Action Step: The "Storm Watch" Drill

You need to trust your sentry before the storm hits. You need to know what the alarm sounds like so you don't sleep through it.

The Scenario: You are at home. A storm front is moving in, or it is simply a Wednesday.

The Drill

1. **Find the Frequency:** Use your radio's scan function to find the strongest NOAA weather channel in your area.

2. **Set Alert:** Go into your menu and enable "WX Alert" (or hold the WX button, depending on the model). The radio should go silent. It is now "Armed."

3. **The Wait:** Leave the radio on the kitchen counter while you cook dinner.

4. **The Test:** Every Wednesday between 11:00 AM and 12:00 PM (local time), the National Weather Service runs a **Weekly Test**. They send the 1050 Hz tone.

5. **The Result:** Your radio should scream like a banshee (a siren tone distinct from normal traffic) and then open the audio channel to the test message.

If it works, you have verified your "Electronic Sentry." Pack that radio for your next trip with the confidence that it will watch the sky while you watch the trail.

Conclusion

Status - Clear

OBJECTIVE: CEMENT THE TRANSFORMATION from "Consumer" to "Commander."

Key Deliverable: A sustainable maintenance ritual to ensure your system works when the world breaks.

The Transformation: From Amazon Box to Airwave Authority

Take a moment. Look at the radio sitting on your desk or mounted in your truck. Pick it up. Feel the weight of it.

Three months ago, or whenever you started this book, that device was just a piece of plastic and circuitry. It was a gadget you bought because you saw a YouTuber use one, or because you had a vague, gnawing anxiety about cell phone coverage in the mountains. It was an object of mystery, shrouded in intimidating acronyms like CTCSS, SWR, and UHF. You likely looked at it with a mix of hope and confusion, wondering if this black box was actually going to save you or just complicate your life.

You were a **Buyer**. You were operating on the consumer premise that simply owning the object would buy you safety. You thought the transaction was the solution.

Today, looking at that same radio, you should see something different. You don't see a gadget. You see a node in a network. You see a tool that extends your influence beyond the horizon. You see the ability to call for help when the sky turns black, to coordinate a rescue when a truck slides off the trail, or to simply tell your kids to come back to camp for dinner without screaming your lungs out. The mystery is gone, replaced by utility.

You are no longer a Buyer. You are an **Operator**.

This is not a semantic distinction. It is a fundamental shift in psychology and responsibility.

- **The Buyer** hopes the radio works. They rely on luck and factory settings.

- **The Operator** *knows* the radio works because they tuned the SWR, programmed the repeater offset, and drilled the PACE plan with their family. They have verified the physics.

- **The Buyer** fears the silence of a dead channel, interpreting it as a failure.

- **The Operator** knows how to scan, how to break squelch, and how to move to high ground to re-establish the link. They view silence as a variable to be solved, not a dead end.

You have adopted the *Visual Command System*. You have rejected the gatekeeping of the "Sad Hams" who want you to memorize circuit diagrams from 1950 before you are allowed to push a button. You have embraced the pragmatic, "if-this-then-that" logic of the field guide. You have learned that wattage is just water pressure, that antennas are just nozzles, and that a radio without a plan is just a paperweight.

The Core Principles: A Final Review

Before we sign off, let's lock in the core doctrines one last time. These are the pillars that will hold your system up when stress and adrenaline try to knock it down.

1. The Hardware is Secondary; The Mindset is Primary

We spent chapters on Baofengs, Midlands, and Wouxuns. We discussed superheterodyne receivers and quarter-wave antennas. But the specific brand doesn't matter as much as the brain behind it. A trained operator with a $30 blister-pack radio who understands line-of-sight is more effective than a confused novice with a $5,000 base station who doesn't know what a repeater offset is. You now know that "the best radio" is not the most expensive one; it is the one you know how to program in the dark, in the rain, with gloves on.

2. Plain Language is Power

We killed the "10-Codes." We buried "Breaker Breaker." We established that sounding like a professional means speaking English. Under stress, the human brain sheds complex codes. If you scream "10-33" at a frightened spouse, they will freeze. If you say "Emergency, I need help," they will act. "Dad to Mom, location?" is infinitely superior to "Unit Alpha requesting 10-20 from Base." Clarity is speed. Speed is safety.

3. PACE is Your Safety Net

Technology fails. Batteries die. Physics wins. You cannot argue with a mountain blocking your signal. You now have a **PACE Plan** (Primary, Alternate, Contingency, Emergency). You know that when Channel 16 fails, you don't panic - you simply execute the next step in the flowchart. You switch to Channel 20. When Channel 20 fails, you hit the Repeater. When the Repeater fails, you blow the whistle or check the GPS. You have replaced panic with a pre-loaded decision tree, allowing you to act while others are paralyzed.

4. Intelligence Over Noise

You learned that the radio is not just for talking; it is for listening. You are the Sentry. You monitor the NOAA weather channels to detect storms before you see clouds. You scan for local traffic to detect logging trucks before you meet them on a blind corner. You use the radio to build a mental map of the battlefield - whether that battlefield is a chaotic trail run or a stormy campsite. Information dominance prevents emergencies.

The Challenge: The Law of Entropy

Here is the hard truth: **Systems decay.**

Nature hates order. The second you finish installing your mobile unit, the universe begins working to dismantle it. If you leave your truck parked in a field for a year, the tires go flat, the battery dies, and mice build nests in the air filter.

Your radio system is no different. It is under constant attack from the elements and time.

- **Lithium-Ion batteries** self-discharge. Their internal chemistry is unstable. If you leave your handhelds in a drawer for six months, they will be dead when the power goes out. Worse, if left at 0%, they may chemically lock and never charge again.

- **Connectors corrode.** That PL-259 connector on your truck's antenna is exposed to rain, mud, road salt, and UV radiation. It will eventually oxidize, creating high resistance that chokes your signal.

- **Repeater details change.** The owner of the "Saddleback 700" might change the privacy tone, move the antenna, or take the system offline entirely. A frequency list from 2023 is a history book, not a field guide.

- **Skills rust.** If you don't use the NATO phonetic alphabet for a year, you will forget whether "B" is Bravo or Baker. You will forget how to unlock the keypad. You will forget which channel is the Alternate.

The "Buyer" puts the radios in a plastic bin, shoves them in the garage, and forgets them until an emergency strikes. Then, in the moment of crisis, they pull out a dead radio with corroded terminals and scream, "Why doesn't this work?!"

The **Operator** fights entropy with a ritual.

The Solution: The Monthly Radio Check

You need a Standard Operating Procedure (SOP) for maintenance. You need a "Radio Shabbat" - a designated time when you inspect and refresh your comms.

I recommend the **First Saturday of the Month**. It takes 15 minutes. It ensures 100% readiness.

The "First Saturday" Checklist

1. The Power Cycle (5 Minutes)

- Gather every handheld radio you own. Dump them on the kitchen table.

- Turn them on. Check the battery voltage indicator. **Remove the battery and inspect for swelling.** If the battery spins like a top on a flat table, it is swollen (spicy pillow). Discard it safely and order a replacement.

- If they are below 80%, put them on the charger.

- *Crucial Step:* If they are fully charged, leave them *on* for an hour to cycle the chemicals, then top them off. Lithium batteries hate sitting at 0%, but they also degrade if left at 100% for years without use. Using them is good for them; it wakes up the ions.

2. The Mobile Inspection (3 Minutes)

- Walk out to your vehicle. Don't just look at it; touch it.

- Grab the antenna. Wiggle it. Is the mount tight? Did the last car wash loosen the set screws? If the antenna wobbles, your ground plane is compromised.

- Check the coax cable. Is it pinched in the door jamb? Is the insulation cracked by the sun?

- Turn the mobile unit on. Check the voltage readout (usually on the screen). Is your truck battery healthy (12.6V engine off, 13.8V engine on)? A radio that dims when you key the mic is warning you of a failing truck battery.

3. The Comms Check (The Family Drill) (5 Minutes) This is the most important part. It keeps your family involved and comfortable with the gear.

- Give a handheld to your spouse or kid. Send them to the backyard, the basement, or the end of the driveway.

- Get in your truck (or use another handheld).

- **Execute the Protocol:**

  - **You:** "Base, this is Mobile. Radio Check."

  - **Them:** "Mobile, Base. Loud and clear."

  - **You:** "Switch to Alternate Channel 20."

  - **Them:** "Switching 20."

  - **You:** (On Ch 20) "Radio Check on Alternate."

  - **Them:** "Loud and clear on 20."

- **Why do this?**

  - It verifies the hardware works on multiple frequencies.

  - It verifies the batteries work under load.

  - It verifies your family remembers how to unlock the keypad and change channels without you standing over their shoulder.

- ○ It normalizes the radio. It stops being a "scary emergency device" and becomes a normal household tool like a toaster or a flashlight.

4. The Weather Sentry (2 Minutes)

- Tune to your local NOAA weather frequency.

- Verify you can hear the robot clearly.

- This confirms your receiver sensitivity is good. If you can't hear the weather station you usually hear, your antenna system has a problem.

5. The "Software" Update (As Needed)

- Once a year (maybe every January), plug your radios into the computer and open CHIRP.

- Check RepeaterBook.com. Have any new repeaters popped up in your area? Have old ones changed tones?

- Update your radio memories. A radio with 5-year-old repeater data is a paperweight.

Final Encouragement: You Are the Hub

We live in a fragile world. We have built a society that relies entirely on a delicate web of cell towers, fiber optic cables, and server farms. When that web works, it is miraculous. But when it breaks - due to a hurricane, a wildfire, a cyberattack, or just a lack of funding in a rural county - it breaks completely.

When the bars on your cell phone disappear, the illusion of safety vanishes with them. For the average person, this is a moment of terror. They feel cut off, alone, and vulnerable.

But you are not the average person anymore.

When the grid goes down, you do not lose your voice. You simply reach for the microphone. You have an independent, physics-based, decentralized communication network that you own, maintain, and command. You are not dependent on a subscription service or a corporation to talk to your family.

You are the **Communications Hub** for your tribe.

- You are the one who will know the weather is turning before the sky goes dark, because you monitored the NOAA alert.

- You are the one who will coordinate the convoy when the lead truck breaks an axle in the dead zone, because you built the PACE plan.

- You are the one who will hear the call for help from a stranger and have the power to answer.

Own that responsibility. It is a heavy weight, but it is a good weight. It is the weight of competence.

Don't let the skills rust. Don't let the batteries die. Teach your kids. Teach your neighbors. Be the "Elmer" (mentor) that the "Sad Hams" refused to be. Share your knowledge freely and build a community that is resilient, capable, and connected.

The airwaves are free. The equipment is ready. The rest is up to you.

Radio Check. Signal: 5 by 5. Status: Clear.

Whiskey Sierra Lima Charlie Two Three Zero

End of Transmission

Build Your Library

The Baofeng Radio Revolution

YOU LIKELY HAVE A UV-5R sitting in a drawer somewhere. Maybe you bought it before you found this book, or maybe you bought a six-pack for the apocalypse. But right now, it's just a confused piece of plastic with too many buttons.

This is the manual that should have come in the box.

While *GMRS Radios: The Visual Command System* taught you the **tactics** of the GMRS service, *The Baofeng Radio Revolution* teaches you the **mechanics** of the world's most popular handheld radio. We skip the engineering theory and give you the field-tested "buttonology" to make this radio work when it matters.

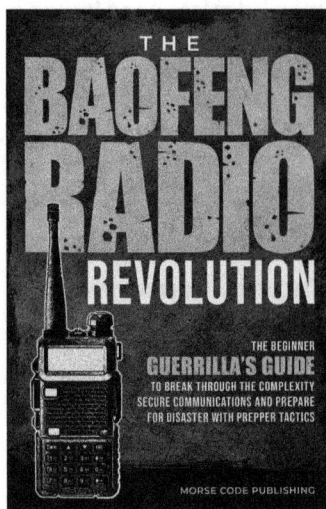

The "Zero-to-Hero" 10-Minute Path: Go from unboxing to your first successful transmission without guessing.

Programming Explained: Learn to program manually in the field (a survival skill) and via CHIRP for speed.

The "Cheat Sheet": Includes a printable 1-page reference guide for your go-bag.

Troubleshooting: Fix the "no audio" and "can't hit repeater" issues that plague new users.

Stop scrubbing through YouTube videos. Get the guide that unlocks the full potential of your hardware.

Ham Radio Technician Class Study Guide

You've mastered GMRS. You understand the power of comms. But you're starting to feel the ceiling. **You want more power, more frequencies, and** the ability to build your own antennas or talk around the world.

You are ready for Amateur Radio (Ham). The only thing standing in your way is The Exam.

Most study guides are dry, boring, and written by engineers for engineers. They make you feel like you need a degree in physics just to push a button. **We built a different kind of guide.**

This book is designed for the person who learns by doing. We break down the FCC exam into bite-sized, plain-English lessons that strip away the confusion and focus on what you actually need to know to pass on the first try.

Why Step Up to Ham?

Higher Power: Legally run up to 1,500 watts (vs 50 watts on GMRS).

Digital Modes: Send email and data over radio waves without the internet.

Don't let "Test Anxiety" keep you on the sidelines. Ace the exam, get your call sign, and unlock the full spectrum of radio capabilities.

Ham Radio General Class License Study Guide

Break the Line-of-Sight Barrier.

GMRS and the Ham Technician license share one major limitation: they are mostly "Line-of-Sight." If a mountain blocks you, you are stuck.

The General Class license changes the laws of physics.

This upgrade gives you access to the HF (High Frequency) bands. These are the frequencies that bounce off the ionosphere, allowing you to talk around the world—or across the country during a regional disaster, without needing a repeater or a satellite. This is the "Holy Grail" of emergency communications.

But with great power comes a harder exam. The General test introduces complex electronics theory and math that stops many operators in their tracks.

We fixed that.

Just like our Technician guide, we strip away the intimidating engineering jargon and focus on the practical concepts you need to pass. We don't just teach you to memorize answers; we teach you how the "magic" of HF radio actually works, using plain English and real-world examples.

Why Go General?

Grid-Down Independence: When local repeaters fail, HF is the only game in town.

Talk to the World: Make contacts in Europe, Asia, and Australia from your backyard.

The Ultimate Prep: Build a station that can reach loved ones continuously, regardless of distance.

Stop playing in the backyard. Get your General ticket and unlock the sky.

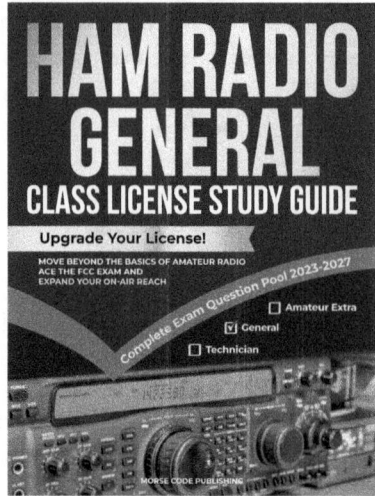

Appendix: The Visual Command Toolkit

Quick Reference Cards

The Visual Command Toolkit: Quick Reference Cards
Print, Laminate, and Keep in Your Kit

GMRS Frequency Chart

Ch	Freq (MHz)	Type	Max Pwr
1	462.5625	Simplex	5W
2	462.5875	Simplex	5W
3	462.6125	Simplex	5W
4	462.6375	Simplex	5W
5	462.6625	Simplex	5W
6	462.6875	Simplex	5W
7	462.7125	Simplex	5W
8-14 (Avoid)	467.5625+	FRS ONLY	0.5W
15	462.5500	Simplex/RPT	50W
16	462.5750	Simplex/RPT	50W
17	462.6000	Simplex/RPT	50W
18	462.6250	Simplex/RPT	50W
19	462.6500	Simplex/RPT	50W
20	462.6750	Simplex/RPT	50W
21	462.7000	Simplex/RPT	50W
22	462.7250	Simplex/RPT	50W

Note: Ch 15-22 are the main GMRS/Repeater Output channels.

NATO Phonetic Alphabet

A - Alpha	N - November
B - Bravo	O - Oscar
C - Charlie	P - Papa
D - Delta	Q - Quebec
E - Echo	R - Romeo
F - Foxtrot	S - Sierra
G - Golf	T - Tango
H - Hotel	U - Uniform
I - India	V - Victor
J - Juliet	W - Whiskey
K - Kilo	X - X-ray
L - Lima	Y - Yankee
M - Mike	Z - Zulu

Numbers:
0 - Zero 1 - Wun 2 - Too
3 - Tree 4 - Fower 5 - Fife
6 - Six 7 - Seven 8 - Ait
9 - Niner

Pro-Words:
OVER: I am done talking, reply expected.
OUT: I am done talking, no reply expected.
COPY: Message received and understood.
SAY AGAIN: Please repeat your last message.

Relevant 10-Codes (GMRS)

Note: Plain language is preferred in GMRS, but these are common shorthand.

- **10-1:** Receiving poorly / Signal weak.
- **10-2:** Receiving well / Signal strong.
- **10-4:** Message received (OK).
- **10-6:** Busy, stand by.
- **10-9:** Repeat message.
- **10-20:** What is your location? ("What's your 20?")
- **10-21:** Call by telephone.
- **10-33: EMERGENCY TRAFFIC ONLY.**
- **10-77:** Negative contact.

Visual Command Rule:
If you are stressed, forget the codes. Speak normally. *"Unit 1, I can't hear you, move to higher ground."* is better than a misused 10-code.

Repeater Quick Guide

The Golden Rule: Offset is ALWAYS +5.000 MHz.

Input (Talk): 467.xxx MHz
↓
Output (Listen): 462.xxx MHz

Troubleshooting Checklist:

1. **Right Channel?** (15-22 only).
2. **Right Tone?** (Check T-CTCS / Transmit Tone).
3. **Offset On?** (Do you see a '+' symbol?).
4. **Wide Band?** (Set to WFM, not NFM).

Field Programming Steps:

1. Go to VFO (Freq Mode).
2. Enter 462.xxx frequency.
3. Menu -> T-CTCS -> Set Tone.
4. Menu -> SPT-D -> Set (+).
5. Menu -> OFFSET -> Set 005.000.

Etiquette:
"[Your Call Sign] monitoring." to announce presence. Don't "Kerchunk" (click mic) without ID.

The Visual Command Toolkit
Repeater Map Log

INSTRUCTIONS: Use this log to record repeaters for your local area or travel route.

- **Source:** Find data on *MyGMRS.com* or local club listings.
- **Input Tone:** The 'Key' your radio transmits to open the repeater (T-CTCS).
- **Output Tone:** The tone the repeater transmits back (R-CTCS). Optional, but filters noise.
- **Offset:** Remember, GMRS Offset is ALWAYS +5.000 MHz.

Location / City	Name	Output Freq	Input Tone	Output Tone	Status	Notes / Range

The Visual Command Toolkit
GMRS Frequency & Bandwidth Chart

LEGEND:

- Shared / Low Power: FRS & GMRS mixed. 5 Watts Max. Narrow or Wide.
- **Restricted (FRS Only):** Low power (0.5W). Narrowband only. **Avoid for tactical use.**
- Main GMRS / High Power: The 'Command Lanes.' 50 Watts. Wideband. Repeater Capable.

CH	Frequency (MHz)	Max Power	Bandwidth	Usage / Notes
1	462.5625	5 Watts	NFM / WFM	Interstitial / Shared
2	462.5875	5 Watts	NFM / WFM	Interstitial / Shared
3	462.6125	5 Watts	NFM / WFM	Interstitial / Shared
4	462.6375	5 Watts	NFM / WFM	Interstitial / Shared
5	462.6625	5 Watts	NFM / WFM	Interstitial / Shared
6	462.6875	5 Watts	NFM / WFM	Interstitial / Shared
7	462.7125	5 Watts	NFM / WFM	Interstitial / Shared
8	467.5625	0.5 Watts	NFM ONLY	FRS ONLY (Avoid)
9	467.5875	0.5 Watts	NFM ONLY	FRS ONLY (Avoid)
10	467.6125	0.5 Watts	NFM ONLY	FRS ONLY (Avoid)
11	467.6375	0.5 Watts	NFM ONLY	FRS ONLY (Avoid)
12	467.6625	0.5 Watts	NFM ONLY	FRS ONLY (Avoid)
13	467.6875	0.5 Watts	NFM ONLY	FRS ONLY (Avoid)
14	467.7125	0.5 Watts	NFM ONLY	FRS ONLY (Avoid)
15	462.5500	50 Watts	WFM (Pref)	GMRS Main / RPT Output
16	462.5750	50 Watts	WFM (Pref)	GMRS Main / RPT Output
17	462.6000	50 Watts	WFM (Pref)	GMRS Main / RPT Output
18	462.6250	50 Watts	WFM (Pref)	GMRS Main / RPT Output
19	462.6500	50 Watts	WFM (Pref)	GMRS Main / RPT Output
20	462.6750	50 Watts	WFM (Pref)	National Emergency / Travel
21	462.7000	50 Watts	WFM (Pref)	GMRS Main / RPT Output
22	462.7250	50 Watts	WFM (Pref)	GMRS Main / RPT Output

Note: Repeaters listen on 467.xxx (Input) and talk on 462.xxx (Output). The Offset is always +5.000 MHz.

Scan to Download Printable Copies.

Glossary

The Visual Command Glossary
100 Essential Terms for the Citizen Operator

1/4 Wave Antenna The most common mobile antenna type. Requires a metal ground plane (roof) to function correctly.

5/8 Wave Antenna A longer antenna that directs signal flatter toward the horizon. Better for distance, worse for hilly terrain.

70cm Band The UHF amateur radio band (420-450 MHz), sitting just below GMRS frequencies.

Airtime The amount of time spent transmitting on a frequency.

AM (Amplitude Modulation) The mode used by CB radio and Aircraft. Susceptible to static.

Analog Traditional radio transmission using voice waves. Fades gracefully with distance.

Antenna Gain A measure of how much an antenna focuses energy. Measured in dB or dBi.

APR (Automatic Packet Reporting). Not used in GMRS, but common in Ham for GPS tracking.

Bandwidth The width of the signal. GMRS uses Wide (20-25 kHz) or Narrow (12.5 kHz).

Base Station A radio installed in a fixed location (home/office), usually with a high external antenna.

Battery Eliminator A fake battery pack that plugs into a car's 12V port to power a handheld indefinitely.

Birdie A false signal generated internally by the radio's own electronics.

Bleed-over Hearing a strong signal from a nearby channel (e.g., hearing Ch 16 while on Ch 15).

BNC A bayonet-style antenna connector. Quick release, common on scanners.

Bubble Pack Derogatory term for cheap, plastic-shell radios sold in pairs at big-box stores.

Channel A pre-set memory slot containing a frequency and settings.

CHIRP Free, open-source software used to program most Chinese radios.

Coax The thick cable connecting a radio to an antenna.

Code Consumer term for CTCSS or DCS privacy tones.

Comms Short for Communications.

Control Operator The license holder responsible for the radio's operation.

Courtesy Tone A beep sent by a repeater to indicate it has reset and is ready for the next person.

Cross-Band Using a radio that can talk on two different bands (e.g., UHF and VHF).

CSQ (Carrier Squelch) 'Open' squelch. No tones used. You hear everything.

CTCSS (Continuous Tone-Coded Squelch System). An analog sub-audible tone used to filter interference.

dB (Decibel) A logarithmic unit used to measure signal strength and antenna gain. 3dB gain = 2x power.

DCS (Digital-Coded Squelch). A digital binary code used to filter interference.

Dead Key Keying the mic without speaking. (See 'Kerchunk').

Desense When a strong nearby signal overpowers a radio receiver, making it deaf to weaker signals.

Direct Talking radio-to-radio without a repeater. (See 'Simplex').

Doubling Two people transmitting at the same time. Result is garbled audio or a loud squeal.

DTMF (Dual Tone Multi Frequency). The 'Touch-Tone' phone sounds used to control repeaters remotely.

Call Sign The alphanumeric ID assigned by the FCC (e.g., WRXP555). Required to identify every 15 mins.

Carrier The silent radio wave transmitted when you hold PTT, before you speak.

CB (Citizens Band) 27 MHz radio service. No license required. Longer range than GMRS in valleys, shorter in cities.

ERP (Effective Radiated Power). The actual power leaving the antenna after cable loss and antenna gain.

Faraday Cage An enclosure (like a metal car body) that blocks radio waves.

FCC Federal Communications Commission. The regulatory body for US radio.

Feedline The cable (coax) running from radio to antenna.

Finals The power amplifier transistors in a radio. They blow if you transmit without an antenna.

FM (Frequency Modulation) The mode used by GMRS. Clearer than AM, but 'capture effect' means strong signals wipe out weak ones.

Frequency The 'address' of the signal, measured in Hertz (Hz) or MegaHertz (MHz).

FRS (Family Radio Service). License-free, low power (2W max), fixed antenna service sharing GMRS frequencies.

Full Quieting A signal so strong there is zero background static.

Gain The ability of an antenna to focus energy.

GMRS (General Mobile Radio Service). Licensed, high power (50W), repeater-capable UHF service.

Dummy Load A device that absorbs radio energy without transmitting it. Used for testing.

Duplex Communication where transmit and receive happen on different frequencies (Repeater operation).

Duty Cycle The percentage of time a radio can transmit without overheating.

DX Long-distance communication (usually via atmospheric skip). Rare in UHF/GMRS.

IP Rating (Ingress Protection). Measures water/dust resistance (e.g., IP67).

Kerchunk Briefly keying a repeater without speaking to see if you can hit it. Considered rude.

Keying Up Pressing the PTT button.

kHz Kilohertz (1,000 Hz).

License The FCC permit required for GMRS. Covers the whole immediate family.

Lid Slang for a bad or rude radio operator.

Line of Sight (LOS) The direct path between antennas. UHF waves essentially stop at the horizon.

Low Band Often refers to VHF Low or HF. Not used in GMRS.

Mag Mount An antenna base using a magnet to stick to a car roof.

MHz Megahertz (1,000,000 Hz).

Mobile Unit A higher-power radio designed to be installed in a vehicle.

Modulation The process of encoding voice onto a radio wave.

Monitor A button that opens the squelch completely to hear weak signals.

MURS (Multi-Use Radio Service). License-free VHF service. Good for woods, but low power.

Ground Plane A metal surface (car roof) required for 1/4 wave antennas to reflect the signal.

Half-Duplex The standard mode of walkie-talkies: You can talk OR listen, but not both at once.

Ham Slang for Amateur Radio Service. Requires a technical exam.

Handheld (HT) A portable walkie-talkie.

Hardline Heavy-duty, rigid coaxial cable used for permanent tower installations to reduce loss.

Harmonics Unwanted signals transmitted at multiples of the main frequency.

High Power Usually 50W for mobile, 5W for handheld.

Hz (Hertz) Cycles per second.

Input Frequency The frequency a repeater listens to (467 MHz range in GMRS).

Interference Unwanted noise or signals disrupting comms.

Intermodulation Interference caused by two signals mixing in a receiver.

Propagation How radio waves travel through the environment.

PTT (Push-to-Talk). The button you press to transmit.

Radio Check Calling out to ask if your signal is being heard.

Range The distance a signal travels. Heavily dependent on height and terrain.

Repeater A device that receives a signal and re-transmits it at higher power/elevation.

RF Radio Frequency.

Narrowband (NFM) A standard using 12.5 kHz bandwidth. Required for FRS channels 8-14.

NMO (New Motorola Mount). The standard screw-on mount for mobile antennas.

NOAA National Oceanic and Atmospheric Administration. Weather radio broadcasts.

Noise Floor The level of background static and interference.

Offset The difference between a repeater's Input and Output frequencies (+5 MHz for GMRS).

Output Frequency The frequency a repeater transmits on (462 MHz range in GMRS).

Over Pro-word meaning 'I am done talking, waiting for reply.'

Part 95 The section of FCC rules governing GMRS and FRS.

PL Tone Motorola trademark for CTCSS.

Prepper Someone preparing for emergencies. A major demographic for GMRS.

Privacy Code Marketing term for CTCSS/DCS filters. Does NOT provide actual privacy.

Tail The moment a repeater keeps transmitting after the user stops talking.

Time-Out Timer (TOT) A setting that stops transmission after a set time (e.g., 60s) to prevent overheating.

Tone Scan A function to search for the CTCSS/DCS tone being used on a busy channel.

Transceiver A device that can both Transmit and Receive.

TSQL (Tone Squelch). Requires a tone to *receive*. Acts as a filter.

Roger Beep A noise sent when PTT is released. Annoying to most pros. Turn it off.

RX Receive.

Scan A mode where the radio cycles through channels looking for traffic.

Scrambler Voice inversion technology.   Illegal on GMRS.

Simplex Direct radio-to-radio communication on the same frequency.

Skip Atmospheric reflection allowing signals to travel hundreds of miles. Rare on UHF.

SMA The small screw-on connector found on most handheld radios.

Squelch A gate setting that silences background static until a strong signal appears.

Squelch Tail The burst of static heard when a transmission ends (before squelch closes).

SWR (Standing Wave Ratio). A measure of how well an antenna is tuned. High SWR damages radios.

TX Transmit.

UHF (Ultra High Frequency).   300 MHz to 3 GHz. GMRS is UHF. Good for penetrating buildings/urban areas.

Unity Gain An antenna with no magnification (0 dB gain). Good for mountainous/hilly terrain.

VHF (Very High Frequency).   30 MHz to 300 MHz. MURS/Marine/Ham. Better for woods/distance than UHF.

VOX (Voice Operated Exchange). Hands-free transmission. Usually unreliable in noisy environments.

Watt Unit of power.

Wideband (WFM) A standard using 25 kHz bandwidth. Preferred for GMRS for better audio quality.

Wouxun A popular Chinese brand known for higher quality GMRS radios than Baofeng.

Yagi A directional antenna that looks like a TV aerial. Focuses signal in one direction.